THE SPOTTER'S GUIDE TO URBAN
ENGINEERING

THE SPOTTER'S GUIDE TO URBAN
ENGINEERING

Infrastructure and Technology
in the Modern Landscape

Claire Barratt & Ian Whitelaw

FIREFLY BOOKS

A FIREFLY BOOK

Published by Firefly Books Ltd. 2011

First printing.

**Publisher Cataloging-in-Publication Data
(U.S.)**

Barratt, Claire.
    The spotter's guide to urban engineering :
infrastructure and technology in the modern
landscape / Claire Barratt and Ian Whitelaw.
[224] p. : col. ill., col. photos. ; cm.
Includes index.
Summary: Examples of urban engineering with
descriptions: what it is, how it works and why it is
there; including: raw materials (mining, agriculture,
waterworks, manufacturing), power (electricity grids,
oil and gas extraction and distribution, and renewable
energy), transport (highways, railways, bridges,
tunnels, canals aviation), communications (telephone,
radio, television, satellites, digital technology) and
waste (sewers, industrial and residential waste
management, recycling).
ISBN-13: 978-1-55407-708-3 (pbk.)
1. Civil engineering. 2. Engineering systems.
I. Whitelaw, Ian. II. Title.
624 dc22   TA145.B3774   2011

**Library and Archives Canada Cataloguing in
Publication**

Barratt, Claire
    The spotter's guide to urban engineering:
infrastructure and technology in the modern
landscape / Claire Barratt and Ian Whitelaw.
Includes index.
ISBN 978-1-55407-708-3
1. Civil engineering. 2. Structural engineering.
I. Whitelaw, Ian, 1953- II. Title.
TA148.B37 2011   624   C2011-904064-6

Published in the United States by
Firefly Books (U.S.) Inc.
P.O. Box 1338, Ellicott Station
Buffalo, New York 14205

Published in Canada by
Firefly Books Ltd.
66 Leek Crescent
Richmond Hill, Ontario L4B 1H1

Conceived, designed and produced by
Quid Publishing
Level 4, Sheridan House
114 Western Road
Hove BN3 1DD
England

Printed in China

I would like to dedicate this book to Julie, to thank her for her support and for keeping me on schedule.

Ian

To Bea and Mirabelle, for all your urban engineering questions.

Claire

CONTENTS

CONTENTS

CHAPTER 4
TRANSPORT

CHAPTER 5
COMMUNICATION

CHAPTER 6
WASTE

INTRODUCTION

Welcome to *The Spotter's Guide to Urban Engineering*, an illustrated identification guide to the structures and technology that surround us in the modern landscape.

Spotters' guides traditionally focus on the natural world and the identification of flora and fauna, but if we are to understand what we see around us today we also need to recognize and understand the landscape's manmade features. We're all experts at using urban engineering. When you grow up with water on tap, paved roads, and garbage collection you don't have to read a manual or pass an exam in order to use it, but as a result we pay very little attention to a whole range of complex and interconnected systems that actually support our way of life.

When flood defenses fail, the consequences can be devastating, as seen in New Orleans in 2005 (see p. 40).

A stone quarry producing aggregrate for the construction industry (see p. 16).

The Spotter's Guide to Urban Engineering is divided into six chapters that cover the main areas of infrastructure—raw materials, water, power, transport, telecommunications, and waste—each of which begins with a brief introduction and a timeline showing the development of the particular area. The various engineering features and structures are then explored, detailing what they are, what they do, how they do it, and, most importantly, how to identify them.

The Spotter's Guide to Urban Engineering provides an introduction to the technology that underpins modern life, throwing light on the anonymous industrial sites, unremarkable roads, strange towers and masts, bewildering electrical installations, and myriad bridge designs we see every day, as well as on the hidden world of pipes, drains, and cables that lies beneath our feet. This book explains the detail behind the design—the why, what, when, and where that answer the fascinating questions you never thought to ask.

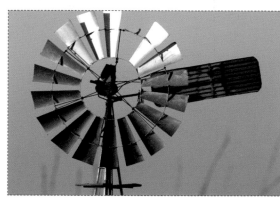

The simple yet effective wind engine has been providing water in arid parts of the world for centuries (see p. 54).

Each entry is fully illustrated, some with extra explanatory diagrams. Famous real-world examples, such as the Millau Viaduct and the Golden Gate Bridge, further illuminate the text.

In Chapter 1, Raw Materials, we start our exploration of infrastructure by examining the extraction, processing, and manufacture of the main materials used in construction, covering topics such as open pit mining and the making of steel and concrete.

Water, the most vital of resources, is the subject of Chapter 2. How is it collected, stored, purified, and delivered to our homes? And how do we protect ourselves against its awesome power?

Increasingly, modern structures are being designed to blend in with the natural environment (see p. 184).

The multilane highway is one of the most striking monuments to the modern urban environment (see p. 110).

Power in all its forms is the subject of Chapter 3. Without electricity or fossil fuels, society in the developed world would simply fall apart, and this chapter looks at how the various sources of energy—including renewable sources that may hold the key to a sustainable future—are harnessed and how energy is distributed to provide power to homes, industry, and transport.

Chapter 4 looks at the crucial infrastructure that supports the various kinds of transport, tracing the historical development of roads and highways, canals, railroads, and rapid transit,

and exploring the facilities and systems that make shipping, cargo handling, and air travel possible.

The subject of Chapter 5 is communications, from billboards and mail services to radio and television, digital telephony, and satellite technology, while in the final chapter we examine the inevitable consequence of all our actions—waste. From industrial flue gases, human sewage, and storm water to garbage and scrap metal, all our unwanted by-products have to be dealt with, and Chapter 6 investigates the engineering solutions that solve the problem.

Waste management is a huge logistical challenge, and one that is constantly evolving (see p. 204).

RAW MATERIALS

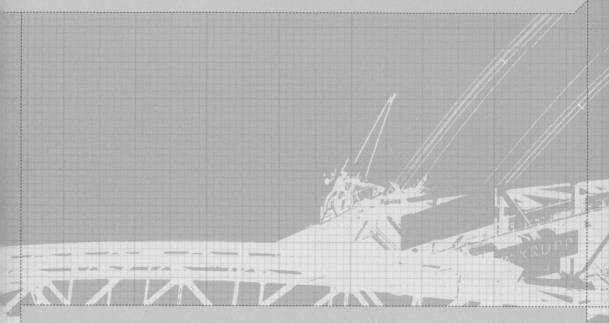

A BRIEF HISTORY OF RAW MATERIALS

40,000 years ago: Humans mine iron oxide (for use as pigment) and Neanderthals mine flint (to make weapons)

11th century BCE: Fired clay bricks in use in Ancient China

5th century BCE: Blast furnace in use in Ancient China for smelting iron ore

1st century BCE: Romans use lime in the preparation of cement, concrete, and underwater concrete

1st century CE: Romans develop large-scale mining techniques

1491: Iron smelting blast furnace introduced to Britain

1627: Explosives first used in underground mining, in Hungary

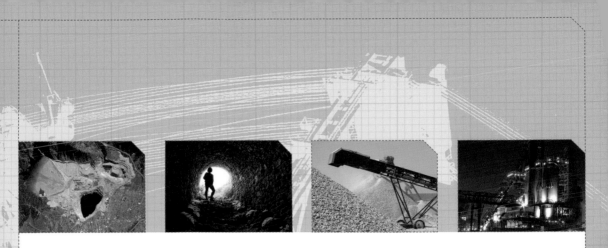

Minerals are naturally occurring substances found in the Earth's crust, and they are essential raw materials in virtually everything that you will find in this book. In this chapter we look at the principal ways of extracting, processing, and using some of the most important minerals.

We begin by exploring the various kinds of mining techniques—from surface mines and quarries to deep underground mines—and the equipment used to extract a range of minerals that includes metal ores (notably iron ore for use in steelmaking), stone for building, crushed rock (aggregate), and coal. We then look at the waste produced by mining, the material that is left behind once the valuable minerals have been extracted and processed.

The second part of this chapter examines the processes that transform minerals into the materials used in construction—the transformation of clay and other minerals into bricks, the manufacture of cement from limestone, the mixing of cement and aggregate to make concrete and a range of concrete products, the creation of road surfacing from asphalt and aggregate, and the smelting of iron ore to make iron and ultimately steel.

1709: Coke replaces charcoal in iron smelting process

1824: Joseph Aspdin invents Portland cement

1858: Henry Bessemer invents the Bessemer converter for steelmaking

1867: Joseph Monier exhibits his newly invented reinforced concrete at the Paris Exposition

1904: John W. Page invents the dragline excavator

1920s: Bucket wheel excavators come into use for open pit and strip mining

1952: Basic oxygen furnace first used for commercial steelmaking

OPEN PIT MINE

WHAT?
A huge manmade hole in the ground from which rock and minerals have been removed.

WHERE?
This kind of mining occurs around the world, but most notably in North and South America, Africa, and Australia.

DIMENSIONS
The largest open pit mine by area is the Hull Rust Mine near Hibbing, Minnesota, which is 5 miles (8 km) long and 2 miles (3.2 km) wide.

The vast majority of the mineral resources that are used to build the infrastructure that we see around us come from the uppermost layers of the Earth's crust and are extracted by surface mining (as opposed to underground mining). The most common form of surface mining, which is also the least expensive and the one that leaves the greatest mark on the landscape, is open pit mining. Many open pit mines have been producing continuously for more than 100 years.

Open pit mining is used when the valuable minerals are close to the surface and the layer of material on top of it, known as the overburden, is not very thick, or where the ground is too soft or too unstable to allow tunneling.

The most common minerals to be mined in this way are ores of metals such as iron, copper, aluminum, gold, and silver, as well as diamonds in Russia and Canada. Because extraction is relatively cheap, mining can be profitable even when the rock contains only a small proportion of the valuable resource.

Digging Down
The process begins with the removal of the topsoil and waste rock that is covering the mineral deposit, using large excavators and trucks to carry the spoil away. In some cases this can take years. All the waste material is deposited around the perimeter of the mine.

Once the mineral-bearing rock is exposed, the pit is worked by blasting and excavating "benches"—terraces on successive levels creating steps down toward the center of the ever-deepening pit. This enables the digging equipment to operate on a level surface and reduces the likelihood that the sides of the pit will subside.

A road is built along one side of the pit so that the extracted material can be hauled out by huge trucks and taken to a processing facility, generally close to the mine. The larger mines use electric shovels that can scoop up almost 100 tons (90 tonnes) at a time, and the trucks can carry up to 350 tons (320 tonnes) of rock.

A Sight to Behold
Although they present a barren, lifeless landscape, the sheer scale of these mines is spectacular, and there are viewing points, museums, and visitor centers at

many mines. These include the US Borax Boron Mine (located on imaginatively named Borax Road, near the town of Boron in California) and the world's largest open pit mine by volume (and the largest manmade excavation), the Kennecott Copper Mine in Bingham Canyon, Utah, which is three-quarters of a mile (1.2 km) deep and covers almost three square miles (7.7 sq km). At some working mines, you can even watch (from a distance) the blasting operations that are used to shatter the rock.

Europe's largest open pit mine is the Corta Atalaya in Andalusia, Spain, which is no longer in use and is partly flooded. Its strange, rocky contours were used as a location in the making of the acclaimed Spanish science-fiction film *PROXIMA*, released in 2007.

Reclamation

Excavation usually continues until the resource is mined out or extraction becomes uneconomical, at which point there are some environmental issues to deal with. Even during the course of mining, water drainage can be a problem and the pits have to be constantly pumped out, as an accumulation of water can dissolve minerals and then carry them into ground water. Some of these huge holes have been lined with clay to prevent leaching, and have then been used as giant landfills for garbage that can eventually be topped with the original topsoil. Others, where contamination is not an issue, have been allowed to become lakes.

TOXIC NIGHTMARE

More than a mile (1.6 km) long and a third of a mile (500 m) deep, the Berkeley Pit in Montana was in operation for almost 30 years, producing copper, gold, and silver until 1982, when the pumps were turned off. Now more than half full of water, it has become a highly toxic lake of dilute sulfuric acid suffused with arsenic, cadmium, pyrite, zinc, and so much copper that the metal is now being recovered commercially from the water. There are fears that the water supply of the town of Butte, just over the rim, might be affected, and the mine has now been added to the Federal Superfund list, earmarked for monitoring and cleanup.

OPEN PIT MINE

Previously excavated benches provide access to the center of the pit and also reduce the risk of subsidence as the mine gets deeper.

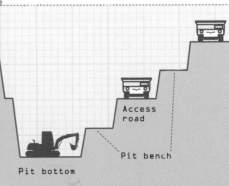

Access road

Pit bench

Pit bottom

STRIP MINE

WHAT?
A large area of shallow surface mining.

WHERE?
Wherever coal deposits are found beneath a thin overburden or under a mountaintop (where tunneling is impossible).

DIMENSIONS
Although they are less deep than open pit mines, some strip mines cover thousands of acres.

Also known as open-cast mining, strip mining is used when the resource— most commonly coal—is found in a relatively shallow seam very close to the surface. Although the popular image of coal mining is of shafts deep underground (see p.18), some 40 percent of the world's coal comes from open mines, and in Australia (New South Wales) and the USA (especially Eastern Kentucky, West Virginia, and Pennsylvania) this method accounts for the majority of coal produced. Strip mining is also practiced in Wales, Germany, and countries in Eastern Europe.

Strip mining is similar to open pit mining in that it begins with the removal of the overburden, but this is taken off in relatively shallow strips using large-scale machinery, exposing the mineral seam, which can then be excavated. Explosives are often used to loosen the overburden and to break up the material to be mined.

On flat terrain, a method called area strip mining is used, and here the overburden is removed in a sequence of parallel rectangular strips, the overburden from each strip being used to fill the previously excavated trench. The last trench is filled with the waste from the first. The area can then be leveled using bulldozers, and the surface can be replanted as a reclamation measure, although the quality of the surface is usually much poorer as the topsoil has now been mixed with waste rock.

Mega Machinery
Strip mining and open pit mining involve the removal of vast quantities of surface material, and some of the largest machines ever built have been developed to achieve this. There are two principal types: dragline excavators and bucket wheel excavators.

The basic design of the dragline excavator has changed little in 100 years, the essential elements being a huge boom, a giant bucket suspended from it by a hoist cable, and a dragline running from the bucket to the body of the excavator. As the bucket is pulled toward the excavator, it scoops up the loosened rock. It is then raised and swung aside to dump the contents of the bucket.

Dragline excavators are so large that they have to be built on site. They are powered by electricity with a direct hookup to the grid, and are able to move very

BIGGEST OF THE BIG

The largest BWE ever built is the Bagger 293, made by the German company TAKRAF. Weighing 15,650 tons (14,200 tonnes), with a boom 720 ft (220 m) long, it works at the Hambach brown coal strip mine in Germany, where it is capable of moving 314,000 cubic yards (240,000 m³) of rock per day. Traveling on 12 caterpillar tracks that spread its enormous weight to avoid damaging the ground, it has a maximum speed of 0.37 mph (0.6 km/h).

DRAGLINE EXCAVATORS

The world's largest working dragline excavator is the Bucyrus-Erie (B-E) 2570WS, known as Ursa Major, located at the Black Thunder coal mine in the Southern Powder River Basin of Wyoming. Weighing 6,700 tons (6,100 tonnes), it has a 360-ft (110-m) boom and its bucket holds 160 cubic yards (122 m³) of material.

slowly across the ground using long pontoons that lift the huge machine and "walk" it forward.

Bucket wheel excavators, or BWEs, are the largest moving terrestrial machines ever built. They have a series of buckets mounted on a large rotating wheel at the end of a boom. The revolving buckets scoop away the rock and empty it onto a conveyor belt behind the wheel that carries it toward the excavator and onto subsequent conveyors that carry the material away. BWEs are used in large-scale surface coal mines and also in Canada's Athabasca tar sands, where bitumen is extracted from the oil-rich sand deposits.

Mountaintop Removal
A variant of strip mining, mountaintop removal involves precisely what the name implies. Where a seam of coal lies beneath the summit of a large hill or mountain, the top of the mountain is blasted apart to expose the coal, and the soil and rock are dumped into the surrounding valleys. This has become the dominant form of mining in West Virginia, and there is widespread concern about its effects—deforestation, the burial of streams, and the long-term destruction of the landscape.

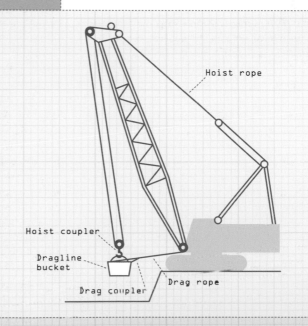

Hoist rope

Hoist coupler

Dragline bucket

Drag coupler

Drag rope

QUARRY

WHAT?
A large open pit from which construction material is mined.

WHERE?
Quarried materials are needed wherever there is construction, and quarries are found in most well-populated areas.

DIMENSIONS
Generally smaller in extent and depth than ore-producing open pit mines, but no less obvious on the landscape.

A quarry is essentially an open pit mine, but the term is applied to operations that produce various kinds of sand and stone for use as building material. As such, quarries are vital for the construction of almost every form of infrastructure mentioned in this book. The mined material does not generally need to be processed (other than being cut, crushed, or graded), it does not usually raise issues of toxicity or contamination, and disused quarries often become picturesque lakes.

Quarries usually look much like open pit ore operations, although they are generally somewhat smaller. The methods of excavation and transport are also similar except that very much less waste material is produced. What makes quarries interesting is the range of essential materials that comes from them, materials that we see all around us every day.

Dimension Stone
Stone that has been quarried in large pieces and then cut to produce regular and specifically sized blocks or tiles is know as dimension stone. The types of stone used are typically granite, marble, limestone, and slate. Stone blocks are rarely used as a building material nowadays due to the cost (although many new buildings are faced with stone veneer), but arch bridges (see p. 148) and the abutments that support other kinds of bridges were commonly built of stone until steel and concrete took over. Roadside curbs

are often made of cut stone, especially granite, and thinner slices are still used today for countertops, floor tiles, and, in the case of slate, roof tiles.

Concrete Ingredients
Portland cement, the binding agent in mortar, grout, and concrete, is made from limestone, clay, and gypsum, all of which come from quarries. Mortar and grout, used to bind together bricks, concrete blocks, or cinder blocks in construction and to fill the joints between them, are a mixture of cement, water, and sand, which also comes from quarries.

Concrete (see p. 26) is made by adding cement and water to a mixture of stone particles of varying sizes, from sand, through gravel and small stones, to large stones and crushed rock. Together these are known as aggregate, and this is by far the most significant product of quarries.

Aggregate

A quarry that is located on bedrock can produce dimension stone or the rock can be crushed to create particles of any size. If, on the other hand, the quarry contains a mixture of sand, gravel, pebbles, stone, and rocks, these can be passed through sieves to separate the different sizes. In either case, the quarry is able to supply whatever size and kind of aggregate is required for the specific purpose—and there are many different purposes besides concrete.

As well as adding strength to concrete, finer grades of aggregate can be mixed with asphalt to form asphalt concrete, the material widely used to build "blacktop" roads (see p. 28). Aggregate can be compacted to form a very stable foundation and it is widely used as the base on which roads are laid. Crushed rock is also used as "track ballast" beneath railroads, providing good drainage, supporting the railroad ties, and preventing the growth of vegetation. Because clay-free aggregate allows water to pass through it, it is also used in applications such as French drains and septic fields. All in all, the worldwide annual consumption of aggregate is estimated to be some 12 billion tons (11 billion tonnes).

WHEN IT'S OVER

Many of the artificial lakes that you see were once quarries. Gravel pits and sand quarries are often located in river valleys and lie below the water table, while clay and stone quarries are generally situated on impermeable bedrock. These naturally fill up with water to form attractive clearwater lakes and find a new lease of life as nature reserves or recreational amenities for water sports such as water skiing, fishing, and swimming. They can be extremely deep and the water tends to be cold, so they need to be treated with caution and respect.

UNDERGROUND MINE

WHAT?
A system of underground shafts and tunnels created to extract minerals from underlying deposits.

WHERE?
Underground mining takes place in most parts of the world.

DIMENSIONS
Mines vary in size from a short single tunnel to gigantic underground systems of interconnected galleries. The world's deepest mine, the Tau Tona gold mine near Carletonville in South Africa, is 2.4 miles (3.86 km) deep and comprises some 500 miles (800 km) of tunnels.

Where mineral deposits occur deep in the ground and the overburden is too thick to make surface mining practical, or where minerals occur as veins in hard rock, underground mining is used to extract them. At the surface you are likely to see machinery that raises the hoist in a shaft mine, or large openings in a hillside from which rail wagons or even large trucks emerge, but the heart of the operation may be hundreds of feet beneath you.

ABANDONED MINES

Working mines have complex ventilation systems to bring in oxygen and to remove potentially lethal gases. They are also often pumped to remove water, and they contain structural supports that need to be maintained, so when the minerals in a mine are exhausted and it is abandoned, it can become an extremely dangerous place. Worldwide, dozens of over-curious people die in abandoned mines every year.

Methods of underground mining vary widely, depending on the minerals being extracted, their distribution, the depth at which they are found and the direction in relation to the surface, and the geology of the area. These factors will determine whether the mine will run vertically, horizontally, or on a slope.

Shaft Mining
When the ore deposits are located directly beneath the ground's surface at depth, a shaft is excavated vertically down to the level of the mineral-bearing seam. Access to the bottom of the shaft is via a cage, or several cages, lowered and raised by means of a hoist. At the top of the shaft you will see a structure called the headframe in which the lifting gear is mounted, either in the form of a motor directly above the shaft or a giant wheel around which the lifting cables, powered by a motor mounted on the ground at one side, are run. Everything going into or coming out of the mine—workers, equipment, and

material—must be transported through the shaft. From the bottom of the shaft, and possibly at several levels within the shaft, horizontal tunnels called galleries or drifts are excavated through the seam.

Drift Mining

When the minerals can be accessed horizontally, for example when they are deep within a hillside, and there is no need for a shaft, the drifts run directly into the deposits. Where possible, the tunnels are excavated at a level slightly lower than that of the ore-bearing seam so that gravity helps in the removal of the material.

Slope Mining

Where the mineral deposit lies at an angle from a suitable point of access, a ramp or decline is excavated at a sufficiently shallow angle to allow wheeled transport to move up and down the slope. If a direct ramp would be too steep, it may spiral down into the ground, possibly around the ore deposit to allow access at several levels. In deep shaft mines, separate levels within the mine are sometimes connected by means of spiral ramps.

Excavation and Removal

Modern underground mining is highly mechanized, using power tools and automated excavating machines that feed the ore onto conveyor belts. There are two principal strategies in underground mining. The most common is "room and pillar mining" in which a seam is excavated using a "continuous miner," a large rotating toothed steel drum that cuts away at the seam and creates a series of rooms separated by pillars that support the rock above. Once the end of the seam is reached, a practice known as "retreat mining" may be used, taking out the pillars as the machinery moves back and allowing the roof to collapse.

Longwall mining involves the removal of an entire seam using a "shearer" equipped with rotating cutting wheels that runs along the face of the seam. The coal or ore falls onto a conveyor that carries it away parallel to the face.

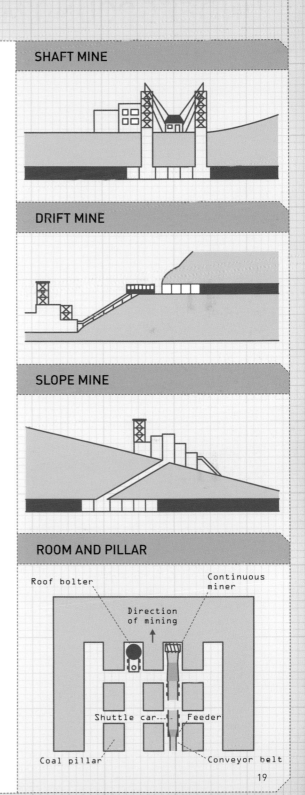

SHAFT MINE

DRIFT MINE

SLOPE MINE

ROOM AND PILLAR

Roof bolter

Continuous miner

Direction of mining

Shuttle car

Feeder

Coal pillar

Conveyor belt

MINE WASTE

WHAT?

Material produced as a by-product of a mining operation, in the form of either waste rock or tailings that result from purifying or processing the extracted minerals.

WHERE?

Wherever there is mining—virtually every kind of mining operation produces some form of mine waste.

DIMENSIONS

The waste from a large-scale operation can cover many square miles and create artificial mountains that transform the landscape.

Having looked at the various kinds of mining and the vast scale on which some of these operations are carried out, it comes as no surprise that mines produce enormous quantities of waste material. Indeed, the first signs of a mine that you will see are often not the machinery nor the gaping hole in the ground but huge hills of earth, rock, and mineral waste.

Mines produce two types of waste: the rock and earth that have to be removed in order to gain access to the minerals; and tailings, the waste that results from the first stages of processing that take place on site to concentrate or purify the minerals ready for shipment. In surface mining, the removal of the overburden creates literally mountains of rock waste, and the primary processing of coal and many mineral ores leads to tailings that contain a wide range of toxic material from the ore itself or from the chemicals used to purify it.

Waste Rock

Mining companies are under an obligation to reclaim the land and restore it to its former state, and in countries with tough regulations the companies have to pledge up to 70 percent of the reclamation costs before they even start mining. In area strip mining, the waste from each strip can be dumped in the previous excavation and the land can then be leveled and, hopefully, replanted. In underground mining it is possible to return some of the waste to the disused workings. In most cases, however, the landscape will have been changed for good. The best that can be hoped for is that the unsightly heaps can be contoured to blend in with the surroundings and to minimize erosion and runoff, covered with topsoil (ideally retained from the original excavation), and then planted with vegetation.

Tailings

The waste that results from concentrating minerals by separating them from the ore generally consists of finely ground rock, water, and chemicals. Solid tailings may be treated in the same way as rock waste, while liquid tailings are pumped to containment areas called tailings ponds where they are expected to dry out. The area then needs to be capped with an impervious material such as clay (so that water doesn't run through the tailings and leach chemicals out of it into groundwater), backfilled with soil, and replanted.

THE LEGACY OF COAL

The carbon-rich tailings from large-scale coal mining operations pose a particular problem. To begin with, the volume of waste is mind-boggling—in China alone, coal mine waste is estimated to be 110 million tons (100 million tonnes) per year, with 1.76 billion tons (1.6 billion tonnes) already produced—creating both an eyesore and a huge risk to groundwater. An even greater problem, however, is the fact that this highly combustible material can, and does, spontaneously ignite, producing unimaginable quantities of greenhouse gases.

Underground coal reserves can also be ignited by surface fires if there is sufficient oxygen present. Many hundreds of coal seams are on fire around the world, and one beneath Burning Mountain in New South Wales, Australia, is thought to have been smoldering for about 6,000 years.

Tailings ponds require careful management.

Water Issues

Runoff and the escape of water from tailing ponds is a major issue, as the tailings often contain toxic chemicals, including arsenic (in gold mine tailings), sulfides that lead to acid runoff, cyanide, radioactive elements such as radium (found in uranium mine tailings), lead, mercury, and other heavy metals. All of these can be highly destructive if they find their way into groundwater or surface streams and rivers.

Dry tailings dust can also blow away from the site and cause pollution over a wide area. Tailings therefore need to be carefully contained; sometimes they require separate treatment plants.

In the Athabasca oil sands (the tar sands, as they are commonly known), the primary processing of the thick crude oil requires huge amounts of water—millions of gallons every year—and the liquid tailings contain bitumen and heavy metals. Although there are plans to use methods of processing that do not require tailing ponds, these lakes of toxic water currently cover an area of some 20 square miles (50 sq km), creating both a wildlife hazard and a management headache.

BRICKWORKS

WHAT?
A factory for the manufacture of bricks for use in construction.

WHERE?
Brickworks are almost always located close to accessible deposits of suitable clay.

DIMENSIONS
A modern industrial brickworks can easily occupy 25 acres (10 hectares) and produce millions of bricks annually.

The humble brick—versatile, relatively inexpensive, and very hard-wearing—reached its peak at the end of the 19th century, when some 15 billion bricks were made annually, and abandoned brickworks with their tall brick-built kiln chimneys can be seen near many major cities. However, bricks are still used extensively for low-rise buildings and for paving, and some 8 billion are made every year in modern, highly automated brickworks that are almost always next to a quarry.

Bricks in the form of blocks of dried mud were in use for house building almost 10,000 years ago and still are used in some parts of the world. The Ancient Chinese discovered the benefits of firing clay bricks to give them strength, and the Romans spread this practice across Europe. Brick became a popular construction material from the Middle Ages on, but the rapid expansion of towns and cities in the 18th and 19th centuries in Britain, the US, and Europe, led to a huge increase in demand. The construction of canals, paved roads, and railroads made it possible to move quantities of bricks over greater distances, but the brickworks that manufactured them were usually built close to the raw materials, and that is still the case today.

Brickmaking
The principal ingredients for brick making are clay, sand, iron oxide, and lime, and most brickworks are situated next to a clay pit that provides not only the clay but often the sand and the iron oxide as well. Rock that contains the perfect combination of clay and sand is known as brickearth. The clay provides a malleable substance that can be molded to shape. The sand, which is silica, melts at high temperatures and bonds the brick together, while the iron oxide gives the brick its red color.

The process begins with the excavation of clay from the open mine, usually in the form of dry, crumbly rock. This is brought to the factory by conveyor belt or truck, and it is then crushed, ground, and screened to remove larger particles before being mixed with the other ingredients in precise proportions. This dry powder is then mixed with water in a pug mill that uses rotating knives to cut and fold the mixture into a homogenous clay dough. The amount of water added depends on which of three methods of brick forming the brickworks uses.

A molding process requires soft, wet clay that can be poured into molds—usually wooden forms sprinkled with sand so that the bricks can be released easily after firing.

Stiffer, drier clay is needed for the pressing method, in which the clay is pushed into a metal die. Bricks made by this method usually have a geometric indentation, called a frog, on one side.

The third method, and the one most commonly used in high-volume mass production, is extrusion. Clay with a medium water content passes through a vacuum chamber to remove any air in it (which might cause the brick to crack when fired), and then into an extruder that pushes the clay out through a shaped nozzle in a continuous strip that is cut into brick-length chunks. Bricks made in this way are often hollow or perforated in cross-section, making them lighter and cheaper without weakening them. Stacked on wheeled metal pallets, or cars, the bricks are then passed through a drying tunnel to remove excess water before being fired to harden them and give them their final appearance.

At one time, bricks were stacked in tall beehive-shaped kilns for firing, but now they pass through a series of chambers in a long, low kiln. The bricks are raised to a temperature of about 2,000°F (1,100°C) using gas or solid fuel, and are then allowed to cool slowly before being inspected, packed for shipment, and dispatched.

THE LARGEST BRICKWORKS

Located at Middle Swan, near Perth, Western Australia, the world's largest single-site brickworks occupies more than 250 acres (100 hectares). Midland Brick produces several hundred million bricks and pavers every year, which it supplies throughout Australia as well as to New Zealand, Japan, and Korea. One of several on the site, the brickworks' newest kiln, installed at a cost of 53 million Australian dollars, has the capacity to produce 50 million bricks a year—enough to build 2,500 houses.

Modern brickmaking is a highly automated process.

CEMENT PLANT

WHAT?

A series of buildings, towers, silos, and conveyors built for the production of cement.

WHERE?

Typically located near an abundant source of good-quality limestone in rural areas. Good transport links (road, rail, water) are essential.

DIMENSIONS

Variable, from a few hundred to a few thousand acres. Plants are usually sized according to output: the largest can produce some 4.4 million tons (4 million tonnes) of cement per year.

Cement is the world's second most consumed substance after water. Combining it with water and sand gives you the mortar that holds bricks together; add various quantities of aggregate (crushed stone) and you get concrete. Heavy, hard, and moldable, it is an engineer's dream. The imposing collection of blocks, stacks, and belts that comprise cement plants loom from a landscape with considerable drama. Individual elements, built on an epic scale and coated in the very dust they are in the business of producing, seem full of purpose, yet the nature of that purpose remains tantalizingly mysterious.

Cement plants, such as this one in Bavaria, are invariably located in regions of limestone.

Cement powder is a combination of differing amounts of chemical elements: calcium, silicon, aluminum, and iron. Calcium-rich limestone forms the main ingredient and this is mined from open cast quarries, usually near the plant and administered by its owners. Once transported there, the limestone, along with other raw materials such as sand, clay, iron ore, shale, and chalk, is then broken down into gravel-sized rocks by large primary and secondary crushers. Each element, depending upon the chemical composition of the final powder, is then proportioned correctly and undergoes further crushing and grinding.

With the more modern "dry" process, the powder (known at this stage as raw meal) passes through the vertical cyclone chambers of a tall preheater tower that is usually the most dominant structure of the entire plant. As the meal falls, it is heated by exit gases from the main kiln. In the older "wet" process, the meal was combined with water to produce a slurry that was then fed into the kiln: this was a more energy-intensive process requiring greater heat that is now out of favor. The kiln lies at the heart of the cementmaking process and is the world's largest piece of moving industrial equipment.

Up to 700 ft (210 m) long and 15 ft (4.5 m) in diameter, the kiln is a long steel tube on a slight incline, lined with firebricks, and makes one to three revolutions per minute. The raw materials, fed into the top end, are heated to a partially molten state by burning oil, coal, or gas. Calcium silicate, cement's primary constituent, forms into marble-sized black lumps called "clinker." The clinker lands on a grate where it is cooled by forced air, and is then passed into the ball mill, a horizontal steel tube filled with steel balls that crush it to a super-fine powder. Once gypsum, which regulates the setting time, has been added, it can be called cement, stored in huge silos, bagged up, and transported.

FACTS AND FIGURES

5%	Global manmade carbon dioxide emissions for which the cement industry is responsible
1,474	Integrated cement production facilities worldwide
1824	The year Joseph Aspdin, a British stonemason, first patented what is now known as cement
2,700°F (1,500°C)	Maximum temperature reached in the kiln
3,900	Acres occupied by the world's largest single-kiln cement plant in Sainte Genevieve, Missouri
150 billion	Grains in 1 lb (500 g) of cement

GLOBAL CEMENT OUTPUT

- 100
- 10
- 1

Cement output in 2004 shown as a percentage of the top producer, China (1,029,217,047 tons, or 933,100,000 tonnes)

CONCRETE PLANT

WHAT?
A central concrete-mixing facility that provides ready-to-use concrete for the construction industry.

WHERE?
Premixed concrete needs to be delivered quickly before it starts to set, so concrete plants are always located in areas with plenty of small-scale construction or close to large projects.

DIMENSIONS
A large concrete plant, with space to store large quantities of aggregate and other materials, can cover several acres. The largest mixers can produce batches of up to 8 cubic yards (6 m³) at a time.

Concrete is the manmade material of our time. It is used worldwide for the construction of foundations, buildings, highways, canals, bridges, railroad ties, harbors, wharves—almost any kind of large-scale infrastructure you can think of. Around the world, more than 20 billion tons (18 billion tonnes) of concrete are used every year—and someone has to mix it.

Given these vast quantities, the process of making concrete is, of course, highly mechanized, and the equipment and materials come together at a concrete, or batching, plant. Here the necessary ingredients—aggregate (or stones, sand, and gravel separately) and cement—are stored in large quantities, together with various additives used in the production of specialty concretes. The plant may be located close to an aggregate quarry to keep transport costs down. The cement, which has to be kept dry, is usually stored in a large hopper located above the heart of the operation, the concrete mixer. Aggregate is normally stored in the open and is brought to the mixer by conveyor belt.

Mixing It Up
A modern batching plant is a high-tech and highly automated operation, using self-cleaning conveyors to transport and distribute materials through the system and computer-controlled devices to measure out precise quantities of the various ingredients by weight or volume. When a batch of concrete is to be mixed, the exact recipe (proportions of cement, aggregate, and water; size of aggregate needed; particular additives) is specified by the customer and the materials are delivered to the concrete mixer, which uses revolving blades or paddles to combine the materials and form a homogenous mix. In certain applications, pigments are added to the mix to create colored concrete.

From the time the water is added, the clock is ticking, and the concrete must be poured on site within 90 minutes. Concrete trucks have to be well maintained to avoid a breakdown that might lead to a drum-full of solid concrete. On large construction projects a temporary batching plant may be set up to produce all the concrete required by the project on site, rather than having ready-mixed concrete delivered, thereby reducing the cost.

STRENGTHENING CONCRETE

Concrete has enormous strength under compression, but has little tensile strength. In many—indeed, most—situations, concrete therefore has to be reinforced with a material that does have strength under tension. Without this additional strength, concrete's usefulness as a construction medium would be very limited and it could not be used to make bridges, beams, floors, or columns, as these would simply crack under tension.

Reinforced Concrete

The reinforcement takes the form of steel reinforcing bars (or "rebar") that form a framework around which the concrete is then poured. This not only enables the concrete to withstand a load under tension, but also ensures that if the concrete were to crack it would at least be held together.

Prestressed Concrete

Even greater tensile strength can be imparted to concrete by casting it around steel "tendons"—rods or cables that are already under tension (or "pre-tensioned"). When the concrete has set and is firmly bonded to the steel, the tension on the tendons is released, compressing the concrete member lengthwise and enabling it to bear an even greater load under tension. This makes longer spans possible.

Concrete can also be strengthened with steel that is "post-tensioned." The concrete is cast around a curved tube with cables running through it. Once the concrete has set, hydraulic jacks are used to stretch these cables tight. The cables are then anchored at each end of the concrete member, the jacks are removed, and the tension is transferred to the concrete.

Precast Concrete

So far we have been considering concrete structures that are being poured on site, and in the case of large structural elements this is the only way it can be done. However, some concrete parts can be brought to the construction site precast—even floor, walls, and beams. Like elements poured on site, precast concrete items can be reinforced or prestressed. Railroad ties are a good example of precast concrete.

HOT MIX ASPHALT PLANT

WHAT?
A facility where asphalt and aggregate are mixed at a raised temperature to produce asphalt cement for road surfacing.

WHERE?
If you see roadmaking equipment laying down "blacktop" road surfacing, you can be sure you're not too far from the facility that mixed the material. There are more than 2,000 such plants in the USA.

DIMENSIONS
Hot mix plants vary in size from relatively small portable units used on site to permanent operations occupying several acres. A large hot mix plant can produce up to 275,000 tons (250,000 tonnes) annually.

Asphalt cement, the material that is used more than any other for road surfacing, hardens when it is cool, and it therefore has to be supplied hot. The operations that mix and supply the material are known as hot mix asphalt (HMA) plants, or simply hot mix plants, and they are readily recognizable by their large silos, drums, storage tanks, hoppers, and conveyor belts, and the huge piles of aggregate waiting to go into the mix.

Highways and roads are built in several layers and, although the pavement (or surface) layer can be made of Portland cement concrete, in the majority of cases the material used is asphalt cement, or "blacktop."

Aggregate of various grades is the principle ingredient of asphalt cement.

Commonly known as just asphalt, asphalt cement consists of aggregate (crushed stone, gravel, and sand) bound together with asphalt (also known as bitumen), a sticky black derivative of crude oil. Asphalt cement production worldwide uses about 88 million tons (80 million tonnes) of asphalt a year.

Hot Mixing
There are two basic types of hot mix plant: batch plants that make asphalt in batches as required; and drum plants that make asphalt continuously and can store the asphalt for several days in heated storage silos. In both cases, the basic plant and the steps in the process are the same.

Aggregate of various sizes is stored on site in large quantities, stockpiled in heaps, silos, or bunkers.

ROAD RESURFACING

When a road is resurfaced, a milling machine scrapes away the old surface material, which can be taken back to the hot mix plant for recycling. The road is then brushed clean by a brooming machine and the surface is coated with a layer of emulsified asphalt to provide a bond. The asphalt cement is then tipped into a hopper on a paving machine and the paver lays down a smooth layer. This is rolled to compact it and, once the asphalt has cooled, the road is ready for use.

From here it is moved by dump truck to cold feed bins, each one containing aggregate of a different size. Adjustable gates and variable speed belts at the bottom of the cold feed bins provide the correct amounts of the various aggregates to a conveyor belt that feeds into the rotating drying drum, where the aggregate is tumbled through hot air to remove moisture.

From Drum to Work Site

At normal outdoor temperatures, asphalt is solid or extremely viscous, but at about 320°F (160°C) it is sufficiently liquid to be mixed with aggregate to form a malleable material that can be spread and compacted on road surfaces.

In a drum plant, heated liquid asphalt is mixed with the hot aggregate in the proportion of around 1:20 in the drum. The hot asphalt cement is then conveyed to an insulated, heated silo for storage. From here it can be discharged into trucks positioned beneath it and taken to the work site. In a batch plant, the dried aggregate is kept warm in hoppers, and only when a batch is required are the aggregate and asphalt mixed together before being discharged directly to the truck. Asphalt cement needs to be warm and pliable to be installed, so hot mix facilities need to be located near paving sites.

Another important element in the hot mix plant is the emission control system, a dust collector that filters out fine sand and dust particles that escape from the drying drum and returns them to the mix.

HOT MIX BATCH PLANT

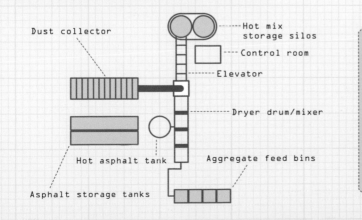

In a batch plant, aggregate from the feed bins is fed into a long rotating drum, where it is heated and dried. Hot asphalt is added and the asphalt cement is mixed. A collector removes dust particles before the mix is taken up to warmed storage silos by elevator. Trucks can drive through under the silos to be filled.

IRON- AND STEELWORKS

WHAT?
Processing plants that extract iron from iron ore and refine iron to produce steel and steel products.

WHERE?
Iron- and steelworks are generally located in regions where iron ore is mined and there is a plentiful supply of coal.

DIMENSIONS
An integrated steel mill, in which all the processing from iron ore through to finished steel products takes place, is the size of a small town, bristling with tall furnaces, chimneys, and conveyors. The Gwangyang Steelworks in Korea has its own power station and produces 18 million tons (16 million tonnes) of steel per year.

Of all the minerals extracted from the ground, iron is the most significant in terms of the engineering and infrastructure that we see around us. Iron and its more refined form, steel, are key components in the construction of railroads, bridges, and anything made using beams, girders, trusses, or reinforced concrete, not to mention cranes, mining equipment, industrial and agricultural machinery, cars, ships, and locomotives. So how does the iron ore in a strip mine become highly refined metal?

Iron is extremely abundant—it makes up about 5 percent of the Earth's crust—but it comes out of the ground in the form of ore mixed with loose earth. The first step in the process is to crush the ore and wash away the lighter earth—this is often done at the mine site.

The ore usually consists of iron oxide—either hematite (Fe_2O_3) or magnetite (Fe_3O_2)—combined with various other minerals such as silicon compounds. The process that extracts the iron from this mixture is called smelting.

Inside a Blast Furnace
Smelting achieves two things: the removal of the oxygen from the iron oxide; and the removal of the other minerals and impurities to leave relatively pure iron. The centerpiece of an iron-production facility is the blast furnace, so called because of the way air is pumped in at the bottom of the furnace. Shaped like a giant chimney, it is lined with heat-resistant brick. The three principal ingredients—ground-up iron ore, coke (a form of charcoal made from coal using a coke oven), and limestone (calcium carbonate, $CaCO_3$)—are poured into the top of the furnace, and air that has been preheated using the waste gases from the furnace is blasted up through the mixture.

As the mixture makes its way down through the furnace—which takes several hours—the coke, which is virtually pure carbon, burns in the hot air to produce a temperature of up to 3,000°F (1,600°C), and carbon monoxide (CO). The carbon monoxide then reacts with the oxygen in the iron oxide to produce carbon dioxide (CO_2) and iron. The calcium in the calcium carbonate combines with the silicates and other minerals in the ore to form slag.

Slag and Iron

These accumulate separately in the bottom of the furnace, with the slag floating on top of the liquid iron, and they are regularly poured out through ports. (A blast furnace runs continuously for several years at a time, as it is not cost-effective to allow the furnace to cool and fire it up again.)

The slag cools to form a hard brittle material that has many uses—as aggregate in concrete and road materials, as railroad ballast, and as a component of phosphate fertilizer. The molten iron can be channeled into a bed of sand, where it cools to become pig iron (so called because the sand mold creates ingots attached to a central bar, like piglets around a sow). Pig iron itself is not very useful, as it contains a high proportion of carbon—up to 5 percent—and is extremely hard and brittle. It can be mixed with slag and hammered to eliminate much of the carbon, creating wrought iron, or it can be melted with scrap iron and other metals and poured into molds to make cast-iron objects, such as manhole covers (see p. 196). Most pig iron, however, goes for further processing to be made into steel (see following page). In an integrated steel mill, the molten iron goes directly into the steel-making process.

BLAST FURNACE

1: Iron ore and limestone
2: Coke
3: Conveyor belt
4: Feeding opening, with a valve that prevents direct contact with the internal parts of the furnace and outdoor air
5: Layer of coke
6: Layer of limestone and iron ore
7: Hot air (around 1,200°C)

8: Slag
9: Liquid pig iron
10: Mixers
11: Tap for pig iron
12: Dust cyclon for removing dust from exhaust gasses before burning them in 13
13: Air heater
14: Smoke outlet (this can be redirected to carbon capture and storage [CCS] tank)

15: Air intake for air heaters
16: Powdered coal
17: Coke oven
18: Coke bin
19: Outlet pipes for blast furnace gas

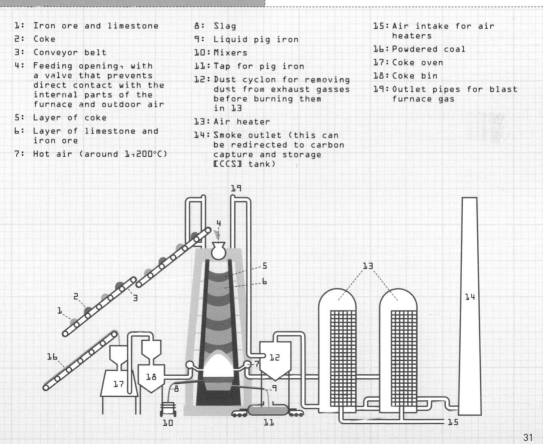

IRON- AND STEELWORKS [CONT'D]

LARGE-SCALE PRODUCTION

The world's largest blast furnace, operated by Shougang Jingtang United Iron and Steel Ltd. in China, stands 164 ft (50 m) tall, has a capacity of 200,000 ft³ (5,500 m³) and is capable of producing 15,400 tons (14,000 tonnes) of iron per day.

From Iron to Steel

The pig iron or cast iron that comes from the blast furnace contains a high proportion of carbon, as well as other impurities. This makes it extremely hard, but not very strong. Wrought iron, which is much purer and contains less than 0.25 percent carbon, is much less brittle and will bend and stretch. A balance between these materials is struck when iron is converted into steel, which is much stronger than pig and wrought iron.

The term steel is a broad one, referring to iron that has a carbon content between these extremes (0.5 to 1.5 percent) and to which other metals may have been added to give it particular properties depending on its purpose. Chromium can be added, for example, to create stainless steel, which resists corrosion. The addition of chromium and molybdenum creates strong, light steel.

The Steel Converter

Since the 1950s, carbon has commonly been removed from the iron using the Linz-Donawitz process in a "basic oxygen furnace," or BOF. The steelmaking furnace, which can hold about 440 tons (400 tonnes) of material, is tilted at an angle in order to "charge" it with one-third recycled scrap steel and two-thirds molten pig iron directly from the blast furnace. A certain amount of limestone is also added. The furnace is then righted and pure oxygen is blown through the mixture, usually from above. The oxygen reacts with the carbon, silicon, and other impurities in the iron, and the oxides form as slag on top of the molten metal. The chemical reaction generates considerable heat, and this is absorbed by the scrap steel, which melts. The process takes about 40 minutes, and the liquid steel can then be tipped out into a preheated container called a ladle. It is then generally cast into large blocks called blooms.

In a rolling mill, which may be a separate facility or part of an integrated steel mill, these blooms are pressed between rollers to produce finished goods such as sheet steel, strips, wire, and structural steel members with a range of profiles.

BASIC OXYGEN FURNACE

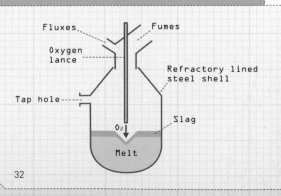

Fluxes
Fumes
Oxygen lance
Refractory lined steel shell
Tap hole
Slag
O_2
Melt

In the basic oxygen furnace, pure oxygen is introduced into the molten mixture via a lance, which is cooled in order to withstand the high temperatures within the furnace. Fluxes are added, and fumes vented, via separate channels. The newly formed steel is "tapped" into a ladle via the tap hole.

ANNUAL STEEL PRODUCTION WORLDWIDE

Because of the importance of steel in the building of infrastructure and consumables such as automobiles, steel production figures are commonly used as an indicator of economic growth. According to the World Steel Association, on whose preliminary figures the table to the right is based, China produced 44.3 percent of the world's steel in 2010, more than five and a half times as much as its nearest rival, Japan. China's proportion of world production has increased steadily in the last decade, as the graph below shows.

**Top 10 steel-producing countries in 2010
(Figures are in millions of metric tons)**

Rank	Country	2010
1	China	626.7
2	Japan	109.6
3	USA	80.6
4	Russia	67.0
5	India	66.8
6	South Korea	58.5
7	Germany	43.8
8	Ukraine	33.6
9	Brazil	32.8
10	Turkey	29.0

ANNUAL CRUDE STEEL PRODUCTION

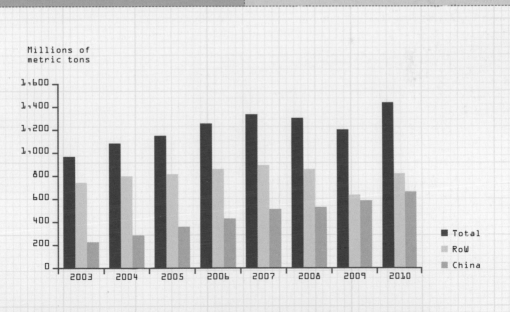

In recent years, annual steel production in China has soared while combined output in the rest of the world has remained largely constant. Between 2003 and 2010, China's output increased by over 300%.

WATER

A BRIEF HISTORY OF WATER

2600 BCE: Early levees are constructed in South Asia

1720 BCE: Tigris River is dammed

1000 BCE: 600 miles of levees are built along the Nile

19 BCE: Eleven aqueducts supply the city of Rome with fresh water

c.250 BCE: Eureka! Archimedes establishes his theory of displacement of fluids

c.100 BCE: The Romans pioneer early water fountains

98 CE: Rome has nine aqueducts feeding 39 monumental fountains and 591 public supplies

1250: The Dutch complete a 78-mile (126-km) dike system

1453: The Acqua Vergine, a ruined Roman aqueduct, is rebuilt, bringing clean water back into Rome

1613: A 40-mile (64-km) channel brings fresh water to the heart of London

The design of every structure has to take into account water: how to control and channel it, and how to keep us protected from it. This chapter examines the engineering behind water management, from the largest structures to the pipes that bring clean drinking water to our homes.

We have engineered levees, dikes, and floodwalls that protect our homes and livelihoods from encroaching seas and oceans, and from the effects of extreme weather conditions. The constant desire of water to move downhill has been harnessed by our distribution networks, which deliver water to areas where it is needed from sources often hundreds of miles away using nothing but gravity.

The functional nature of water-management systems is often disguised within architecturally interesting structures, from the classical beauty of perfectly proportioned Roman aqueducts to the comedy appendages added to American water tanks. Nowhere is this more evident than in the lavish fountains that have graced private homes and public spaces for millennia.

1641: The Shalimar Gardens in Lahore, modern-day Pakistan, contains 410 fountains and pools

1666: The UK's poor waterpump technology cannot prevent the Great Fire destroying much of London

1721: British Engineer Richard Newsham invents the first proper fire engine

1891: Ozone treatment successfully kills bacteria in water

1906: France commissions the first ozone plant to treat water

1919: A formula is established for the chlorination of drinking water

1936: The Hoover Dam is completed

1960s: Kuwait begins using seawater desalination techniques

1970s: The Aswan Dam construction is completed controlling the Nile

2011: China's Three Gorges Hydrolectric Dam nears completion. It will be the world's largest manmade structure

RESERVOIR

WHAT?
A natural or artificial pond or lake used for all a region's water regulation and storage needs.

WHERE?
Worldwide: they are an important means of storing water during times of abundant rainfall, and providing a rationed amount during times of drought.

DIMENSIONS
The world's largest in terms of water volume is Lake Kariba, between Zambia and Zimbabwe, which contains 43 cubic miles (180 km³) of water. The largest in terms of surface area is Lake Volta, in Ghana, which covers an amazing 3,275 square miles (8,482 km²).

Keeping water under control is the most important engineering task. Reservoirs provide several benefits, depending on their location, although engineering works on this scale come at a cost.

Storage and regulation on a large scale may offer a number of benefits: the supply of water for people, industry, and crop irrigation; the generation of electricity; and the management of large rivers. Reservoirs can offer substantial storage capacity, which can help to prevent flash flooding and offers a haven to wildlife.

The Gibson reservoir in Montana, USA.

Valley Dammed Reservoir
By allowing the topography of the site to do most of the work, large-scale reservoirs can be made with relatively little construction. The sides of a valley act as walls and, by building a dam across a narrow downstream section, the natural water level is raised, which creates a huge storage volume and allows regulation of the water supply downstream. This type of construction is perfect for combining water storage with the generation of electricity.

Bank-side Reservoir
These are built for the provision of drinking water where the original supply source may be erratic or polluted. Created by digging down and building up walls, these waterproof tanks are often created with puddled clay. The water is pumped or siphoned from nearby rivers and often stored for several months, allowing natural biological processes to reduce contaminants and the cloudy nature of river water.

The Queen Mother Reservoir in the UK, which stands 45 ft (14 m) above the surrounding fields and motorways, is clearly visible from the air. The sailboats skimming across the 700-acre (300-hectare) reservoir and the sheep munching the grass-covered earth banks contrast strongly with the roar of the nearby trains, planes, and automobiles.

Service Reservoirs

Service reservoirs store fully treated drinking water. Water towers are one type of reservoir, but others are hidden entirely underground. Nineteenth-century reservoirs used brick for the floor, walls, and columns, which supported arches and a roof (both also made with bricks). All that is visible from above is a lawn and a restricted entrance. Munich in Germany has spectacular concrete tanks that seem to go on forever, with roofs supported by pillars that are shaped like cocktail glasses.

Reservoirs are impressive structural achievements. The scale of engineering involved in safely storing large quantities of water should not be underestimated; these structures are built to last and need very little maintenance over their lifetime.

UNDERGROUND CISTERNS

Istanbul stands on the remains of settlements dating back to 660 BCE, which include hundreds of ancient water cisterns. While many are plain, others display architecture fit for a sultan's palace. These underground storage chambers, which held water for times of drought and siege, were filled via aqueducts that stretched to the outskirts of the city. The largest and most impressive contain ornately carved columns reused from even older buildings. The Basilica Cistern built in the 6th century contains 336 marble columns and covers an area of 100,000 ft² (9,800 m²).

HOW A RESERVOIR WORKS

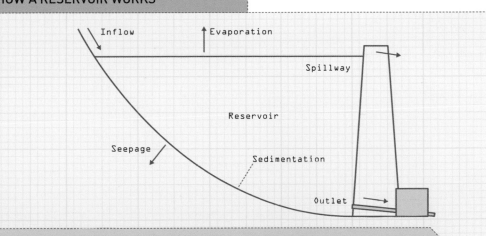

Water enters the reservoir as rainfall and, either directly or indirectly, from a river. Output usage has to be balanced against water loss through seepage and evaporation. If the reservoir gets too full, excess water is discharged over a spillway.

DAM

WHAT?

A barrier to the natural flow of water that is strong enough to keep it stored, but can also release it in a controlled manner when needed.

WHERE?

Dams are ideally situated in narrow sections of deep river valleys, where the sides act as "walls" and thus help to reduce construction costs. There are over 800,000 dams worldwide.

DIMENSIONS

There are currently more than 40,000 dams over 50 ft (15 m) high, including the Three Gorges Dam on the Yangtze River in China, which is five times bigger than the Hoover Dam in the USA.

As an urban spotter, you may have to venture out into the countryside to observe a dam in action. Classification takes into account their width and height, their structure, and their specific purpose.

Timber dams work very well for beavers and also for humankind, but they are limited in lifespan and size. Steel dams were briefly trialed, but masonry, earthworks, and concrete offered the best solution in terms of flexibility and cost. The big dams use concrete in all its forms—mass, reinforced, or prestressed. Asphalt-concrete is particularly popular in seismically active regions due to its flexibility.

Construction of the arch Amagase dam, Japan, was completed in 1964.

Arch Dam

Shaped like a beaver's timber dam, the curve of the arch holds back the water in the reservoir. Arch dams are elegant structures that use a minimum amount of material to resist large forces. The Hoover Dam in Arizona, USA, is a famous example.

Buttress Dam

These can be flat or curved and generally use reinforced concrete to create a characteristic series of supports on the downstream side that braces the dam wall. Examples include the Coolidge Dam in Arizona, USA, and the Clywedog Dam in Wales.

Embankment Dam

Very common in the USA, embankment dams work like gravity dams (see below). They resist the force of water by efficiently using earth and rock to provide weight while a waterproof core prevents the water percolating through the structure. Examples include the Aswan Dam in Egypt.

ARCH DAM

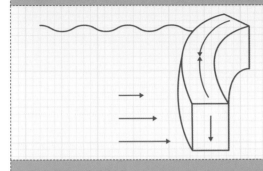

Gravity Dam

Gravity dams are very heavy. They resist the pressure of the water through sheer mass. They need an awful lot of concrete, far more than the thinner arch and buttress dams, and are consequently far more expensive. To the general public, however, their solid and dependable appearance provides a great deal of reassurance. The Grand Coulee Dam in Washington, USA, is an example of a gravity dam.

BUTTRESS DAM

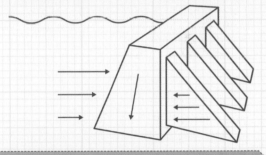

DAM FAILURE

Dam failures can be spectacular and catastrophic, particularly if the structure is breached or significantly damaged. A successfully functioning dam (which allows the engineer to sleep peacefully at night) is one that shows no sign of movement. Routine monitoring is carried out of drains around the dam in order to identify seepage so that action can be taken to address problems before they escalate.

Due to the massive impact that a dam failure will have on the downstream population and environment, dams are considered to be "installation[s] containing dangerous forces." They are protected under international humanitarian law, which states that they must not be targeted during war. The British Royal Air Force "dambusters" had famously attacked dams in Germany during World War II.

EMBANKMENT DAM

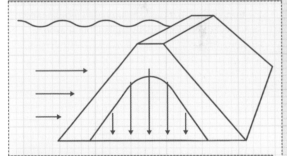

GRAVITY DAM

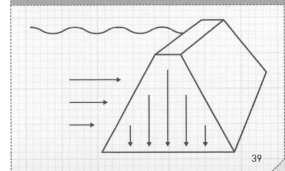

LEVEE, DIKE, AND FLOODWALL

WHAT?

A wall or bank constructed to protect land and buildings from flooding, usually covered in grass in an attempt to make it look less conspicuous.

WHERE?

Flat low-lying land with an unpredictable sea not too far away.

DIMENSIONS

Wide enough to travel along, high enough to resist the sea in even the most extreme conditions, and very long.

The fertile, flat land next to a river or sea may look like a pleasant place to start a new settlement, but when the storms come you may find yourself underwater. Like King Canute, we have been trying to hold back the sea for hundreds of years by building long banks and walls. When the sea is calm they do their job with ease, but when a storm arrives and the water level surges upward, this structural engineering is tested to the limit.

A levee is an embankment designed primarily to provide flood protection from seasonal high water and that is therefore subject to water loading for periods of only a few days or weeks a year. A floodwall is effectively a one-sided levee, providing protection from one direction only. Dikes (also known as dykes) are similar, but they do their job all the time, not just during floods, because the area they guard is below the water level. In many cases, construction of a levee is simple. Impermeable earth, such as clay, is piled up into a mound with a wide base 10–30 ft (3–10 m) high. The higher a levee is, the wider it will be.

Dutch Knowledge

Due to the fact that much of their country lies close to or below sea level, the Dutch are recognized as being the best flood defense engineers in the world. There are many hundreds of kilometers of dikes across the Netherlands protecting land reclaimed from the North Sea. These dikes typically appear as grassy banks with water on one side and marshland on the other. Their exposed location has made them ideal for supporting long lines of turbines providing nearly 10 percent of the nation's power. You can also spot lines of windmills on dikes too, but their job is to pump water from behind the dike back into the sea.

An Ancient Practice

Levees can also be spotted along riverbanks providing defense from surges of high water during periods of very high rainfall. There is evidence of levees being built over 3,000 years ago by the ancient Egyptians along the Nile and by the Harappans along the River Indus in what is now Pakistan and northern India.

Sometimes the most striking feature in a long, uniform levee is the floodgate. These are used to allow a controlled flow of water through and can be large and impressive feats of engineering.

WATER

LEVEE BREACH

Something you hope never to spot first hand, a levee breach occurs when part of a levee, or dike, breaks away, allowing the water it was holding back to flood through. The most recent devastating example occurred in August 2005, when Hurricane Katrina caused the levees and floodwalls protecting New Orleans to breach in more than 50 places, submerging most of the city in water from the Gulf of Mexico. While some of the flooding was caused by "overtopping," where water surges over the top of the barrier, the majority was caused by structural failure. This led to severe criticism of the authorities and engineers responsible for the levees' construction.

Parts of the Netherlands have suffered numerous dike breaches caused by storms in the North Sea. The worst of these occurred between the 15th and 18th centuries and killed thousands of people.

LEVEE CROSS-SECTION

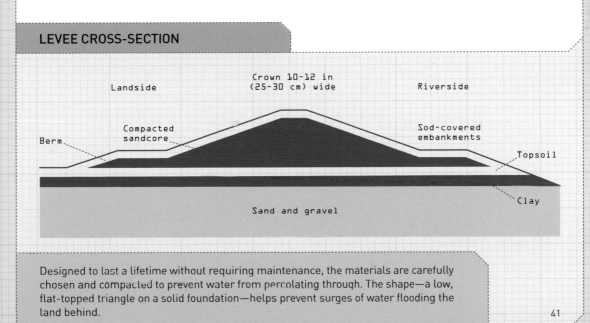

Crown 10-12 in (25-30 cm) wide

Landside

Riverside

Berm

Compacted sandcore

Sod-covered embankments

Topsoil

Clay

Sand and gravel

Designed to last a lifetime without requiring maintenance, the materials are carefully chosen and compacted to prevent water from percolating through. The shape—a low, flat-topped triangle on a solid foundation—helps prevent surges of water flooding the land behind.

AQUEDUCT

WHAT?
A system of canals, pipes, tunnels, and channels that redirects untreated water to areas that need it.

WHERE?
If still entirely intact, one end of the aqueduct will be at a lake or reservoir and the other will be close to a town or city, or emptying onto agricultural land.

DIMENSIONS
Often, all that remains of an ancient aqueduct is a short, narrow stone channel. The modern California aqueduct, however, stretches for 336 miles (540 km) with channels 40 ft (12 m) wide and 30 ft (10 m) deep.

We use a lot of water in our everyday lives, for drinking, washing, and growing our crops. If we are lucky, there is plenty nearby. If not, it has to be transported from elsewhere. As populations grow, the demand for water increases and areas that once had sufficient water supplies find themselves needing more. This problem has existed for thousands of years and one impressive solution is the aqueduct.

Aqueducts are a stunning testament to human ingenuity.

Natural Power
Most aqueducts make use of gravity. A slight gradient over the length of the aqueduct allows the water to flow from source to destination without requiring any additional power. If there is insufficient gradient over the full distance, modern aqueducts use pumping stations to raise the level of the water in a step before allowing gravity to take over again. The reliance on gravity means that careful planning is required when deciding on the path an aqueduct should follow. When the path meets a valley or ravine, a bridge is built, as shown in the image (left).

As with all pipes and channels, an aqueduct must not leak. Just like those built by the Romans, modern aqueduct channels are made from concrete. However, the lead pipes used by the Romans have been replaced by cast concrete tubes. Much of

ANCIENT AQUEDUCT BRIDGES

It is a reflection of the importance placed on aqueducts in ancient society that many still survive. The most striking and impressive parts of these aqueducts are the bridges, spanning valleys and rivers, or majestically crossing towns and cities, undaunted by the rise of modern buildings around them.

These striking examples of ancient engineering have become popular tourist attractions in many parts of the world. Ancient Roman aqueduct bridges can be seen in Italy, France, Portugal, Spain, Greece, and Turkey.

The Pont du Gard in southern France stands 160 ft (50 m) over the River Gard and is the tallest surviving Roman aqueduct. Equally well preserved is the aqueduct in Segovia, Spain. Also a Roman aqueduct bridge, it stands boldly above the city with a total of 167 single and double arches carrying a channel nearly 98 ft (30 m) above the ground. More recently built, but equally inspiring, aqueduct bridges can be found across the world, such as the 18th-century aqueduct in Querétaro, Mexico.

an aqueduct, ancient or modern, is underground. This minimizes accidental or intentional damage, contamination, and, in hot countries, loss of water through evaporation.

Aqueduct, Not Viaduct

Do not confuse aqueducts with viaducts. A viaduct is a bridge composed of spans similar to an aqueduct, but carrying rail or other transport. Occasionally, the two are combined, with the railway line laid above and the water channel below, or the two side by side.

The term "aqueduct" is also applied to bridges that carry navigable water channels. These are larger than water-supply aqueducts and became an essential part of the canal network developed in Europe during the 18th and 19th centuries.

The Pont du Gard, near Remoulins in southern France.

WATER TREATMENT PLANTS

WHAT?
Treating fresh water from rivers, springs, and underground aquifers to make it safe to drink. Treatment plants are crucial in preventing outbreaks of cholera, typhoid, and other deadly waterborne diseases that have killed so many people around the world.

WHERE?
If you were to follow the water pipe back from your house, you would eventually find the storage reservoir, and after that the treatment works.

DIMENSIONS
Municipal water treatment plants cover several acres and often have underground reservoirs.

Water treatment plants perform a number of functions. They include adjusting the pH level, filtering out solids, and disinfecting the water to remove harmful elements such as bacteria. Flocculation sounds and looks horrible, but scraping a layer of scum from the water surface helps to make the water in your glass clear and colorless. The system is designed to exploit gravity wherever possible, although at some stages a pump is used to provide pressure.

Architectural Notes
When spotting your local water treatment plant it is useful to know when it was built, as the site will reflect the architectural trends of the age. A treatment works from the 19th century—there are many still in use—is a grand collection of buildings. The site will have been conceived of as a whole and there is a sense of unity with the tanks, buildings, and even the surrounding wall showing the hand of the same architect. Steam-driven water-pumping stations will have tall chimneys with polychromatic brickwork and elaborate flourishes, such as marbled interior walls with heavy wooden doors that have brass fittings. Rather than replace them with something smaller, many sites have retained these huge buildings, removing the large steam pumps and installing

an electric equivalent that occupies a small area of the floor space.

As the technology of water treatment became smaller and simpler, so did the design of new treatment plants. An unfussy, utilitarian approach characterizes the architecture of plants built in the early decades of the 20th century. In the 1960s, a less aesthetically appealing, "brutalist" architecture became fashionable, which is evident in plants built at that time. Bare concrete contrasted with carefully landscaped grounds, and grass was used to cover extensive underground reservoirs.

WATER

WHY DOES TREATED WATER SOMETIMES TASTE OF CHLORINE?

Adding disinfectant chemicals is the last stage in purifying drinking water. This kills off harmful bacteria that have made it through the filtration process. The most common method involves the addition of a chlorine compound, which rapidly kills many harmful microorganisms thanks to it being a powerful oxidant. In order to guarantee consistently safe, drinkable water, a small amount of the disinfecting agent is present throughout the distribution system, where water may remain for days before finally reaching your home. A residual amount of chlorine will be present in drinking water, giving off an odor that many will be familiar with. This can be alleviated simply by allowing the water to stand in a clean jug for a short period of time.

THE WATER SYSTEM

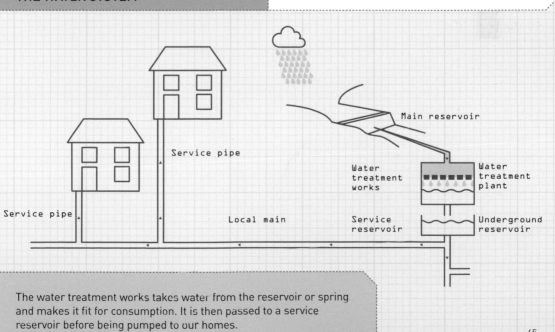

The water treatment works takes water from the reservoir or spring and makes it fit for consumption. It is then passed to a service reservoir before being pumped to our homes.

WATER TOWER

WHAT?

An elevated storage tank that uses gravity to send water flowing through the pipe system to your faucet without direct use of a pump. This system provides water at peak times; the tank is refilled gradually by pump when demand is lower.

WHERE?

If you live in a flat area full of small towns you'll see water towers everywhere; in areas with more bumpy terrain, a simple tank at the highest spot will suffice. Small towers may supply individual buildings or farms.

DIMENSIONS

A small tower will be 20 ft (6 m) high and 13 ft (4 m) across; a standard one will be around 130 ft (40 m) high and hold as much as a million gallons—roughly a day's worth of water for the area it serves.

The universal need for water and the effectiveness of the water tower in providing it on demand, even during power cuts, means the towers will be a distinctive feature in our landscape for years to come.

An ornate water tower at al-Khobar, Saudi Arabia.

The water tower requires an elevated, sealed tank with a pipe going in and one coming out, and a couple of non-return valves along with access for maintenance. These engineering requirements can be satisfied by a bewildering array of materials, shapes, and cladding, although beneath the fancy architectural flourishes it is just a tank of water.

Camouflaged

Masters of disguise, water towers encourage a sense of playfulness lacking in most engineering. Tanks have been painted or clad to look like a giant ear of corn (Rochester, Minnesota), a ketchup bottle (Collinsville, Illinois), and a giant peach (Gaffney, South Carolina). The residents of Thorpeness in the UK were served by a tank disguised as a house floating 70 ft (20 m) above the surrounding trees. Bizarrely, it has now been converted into a house.

The large brick towers of the 19th century have given way to steel and reinforced or prestressed

WATER

WATER TOWERS ON THE NEW YORK SKYLINE

In the 19th century, New York required that all buildings higher than six stories have a storage tank on the roof so as not to have to install regular mains water pipes. The tanks contained sufficient water for the occupants' needs and for use if fire should break out. Barrel makers turned their skills to providing the large wooden water tanks, which are still in service today. The wooden staves are held in place with cables. The barrels will leak at first, until they swell and form a watertight wall that doesn't require sealant.

Some New York architects hide the compulsory water tank behind the building facade or enclose it in an unobtrusive cladding. Others embrace their appearance, leaving them soaring above the rooftops on their utilitarian steel legs.

concrete. The original modernist concrete "mushroom" tank was built in 1958 in Örebro, Sweden. In stark contrast to the local wooden houses, it has a lookout platform and restaurant above the actual water tank. The concrete form has now been copied many times over and is a common sight in continental Europe. It has also spread as far afield as Saudi Arabia and Kuwait.

Backup
Water tanks have to be able to supply emergency mains-pressure water to the community they serve for at least a day, giving time for engineers to reinstate the normal pumped supply. To achieve this, they must be big (the largest holds 1.2 million gallons, or 5.6 million liters) and elevated. Height can be achieved either by building them on high ground or by building them tall. For gravity to provide mains-pressure water, an elevation of about 100 ft (30 m) is required.

The water tower in Rochester, Minnesota.

WATER DISTRIBUTION NETWORK

WHAT?
The pipes that transport water from the treatment plant to homes and factories—cleanly, safely, and at the turn of a faucet.

WHERE?
Mostly underground, but rising above ground to attach to a faucet, hydrant, valve, or fountain.

DIMENSIONS
When they enter our homes, pipes are only 1–1.25 in (25–30 mm) in diameter, but the London Water Ring Main is 8 ft (2.5 m).

While aqueducts transport untreated water from a reservoir or other source to the treatment plant, the water distribution system finishes the job, delivering clean, treated water to our homes and businesses. Clean water is the lifeblood of our towns and cities, and a great deal of urban engineering goes into ensuring that its distribution is safe, efficient, and reliable.

The History of Pipes
The basis of a water distribution system is the pipe. The Romans used lead pipes due to lead's malleability, which made shaping it easier. Lead water pipes can still be found in old houses in Europe, although they are rapidly being replaced because of health concerns. Wooden pipes made from hollowed-out logs were used in Europe during the 16th and 17th centuries, after which iron and copper were used. Although more expensive,

COMMON SUPPLY PIPES (CSP)

The mains pipe is separated from the domestic branch pipe by a stop tap that cuts off supply to the properties. Once inside the property boundary, responsibility for maintenance lies with the land owner, not the distribution provider.

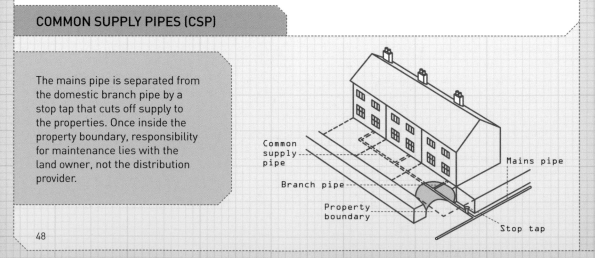

Common supply pipe

Mains pipe

Branch pipe

Property boundary

Stop tap

copper has the advantage of being more corrosion-resistant than iron and is still extensively used in domestic systems today. Larger pipes are made from concrete or plastic. Concrete pipes are stronger in large diameters, but they have limited length due to their weight and lack of flexibility. Plastic pipes are resistant to corrosion and can be used in longer lengths, minimizing the risk of leaky joints. They also have a smooth inside surface, allowing increased water flow rates. As most of the water distribution network remains underground, the best chance you'll get to spot it is when pipes are laid, maintained, or replaced. This is not as rare as it sounds, particularly in cities with aging networks such as London.

Replacing Water Pipes

Nearly half of London's thousands of miles of water mains are more than 100 years old, with some sections as old as 150. These aged cast-iron pipes are more prone to leaks and burst more frequently than modern plastic pipes and are therefore being slowly replaced. Replacement of a water main can be carried out by drilling out the old pipe, drilling a new hole, slotting a new pipe of smaller diameter in the old pipe, or simply digging a trench. The trench option is avoided as much as possible as it causes major disruption to road users. When working without a trench, small access holes are dug about 330 ft (100 m) apart and a steel rod is pushed down the pipe. The new pipe is then pulled back through the pipe or hole fed from a large reel to minimize the number of joints.

Water Pressure

Maintaining water pressure is essential if the distribution network is to function efficiently and without problems. This is achieved either through the use of pumps or establishing a gradient along which the water can flow. The latter is usually achieved by using water towers. Valves regularly spaced along the network provide pressure regulation and help prevent backflow, which can cause contamination. Access points to these valves can provide another way of spotting the location of water supply pipes. You may spot manholes for underground access, or smaller

FIRE HYDRANTS

The most prominent feature of a water supply system in the urban landscape is the fire hydrant, which is used primarily by firefighters to tap into the municipal supply. A hose is attached to the fire hydrant and a valve is opened, producing a powerful flow of water at mains pressure. A fire-engine pump can be used to further boost the flow or allow a number of hoses to work from one hydrant.

To restrict access to hydrants, special tools are generally required to open them, but permission can be obtained by contractors, street sweepers, and even steam traction engine drivers. In many areas, swimming pools may be filled via a fire hydrant on the understanding that if the need should arise the fire brigade may drain the pool dry.

metal plates for surface valve access, bearing an indication that a water supply is below. Each house will have a stopcock to cut off supply under such a metal plate located just outside the property.

FOUNTAIN

WHAT?
Water pouring into a basin to supply drinking water or spouting into the air for dramatic or decorative effect.

WHERE?
Worldwide—everyone loves a fountain. The original function was to provide water for local residents, but the rich and powerful took them to a whole new level.

DIMENSIONS
From the humble village square to lavish private gardens, fountains have practical and decorative value. The King Fahd's Fountain is among the tallest in the world; it runs continuously at a height of 1,000 ft (300 m) over the Red Sea.

The public water fountain has for centuries been a focal point for local gossips as well as providing an essential source of drinking water. Fountains have also been a popular plaything of the rich and powerful. What better way of demonstrating your wealth than building a bigger and more expensive fountain than your neighbor?

The El Alamein Memorial Fountain in Sydney, Australia.

A Symbolic Monument
Water features are one of the most notable parts of our public spaces; they can honor individuals or events, provide a welcome cool spot on a hot day, or just bring a smile to your face thanks to a fun design. A source of water higher than the fountain itself can provide a simple flow. The greater the height differences between source and spout, the higher the pressure and the more spectacular the output. The fountains of St Peter's Square, built in 1614 in Rome, were fed from an aqueduct 265 ft (81 m) above sea level; they could shoot water 20 ft (6 m) up into the air.

Fountains have been used with sculpture in Italy and France for hundreds of years, bringing a sense of movement, peaceful or turbulent, to the marble statues beneath. These fountains are not about shooting the largest jet possible into the sky; instead

MUSICAL FOUNTAINS

Whether it's the latest Las Vegas extravaganza, a show of wealth in Dubai, or a huge fireworks and water show to finish off a long day at Disney World, water has been used for centuries to create spectacle on a grand scale.

Every great exhibition between 1850 and 1950 had to have a variation on the water show. Philadelphia in 1876 introduced fountains illuminated by gas lights; by 1884 there were electric lights in London, and variations followed with music and lights all carefully orchestrated for our enjoyment.

Louis XIV of France has a strong claim to having created the modern musical fountain by staging spectacles in the Garden of Versailles during the 1680s, complete with music and fireworks to accompany the flow of the fountains.

they use the often limited water supply to great effect, producing cascades that please the eye.

Philanthropic Gesture
As well as fountains built purely for display purposes, there are others that were built to provide an important source of water for the local population. Many drinking fountains were set up by wealthy individuals to provide free drinking water to the poor and discourage them from drinking alcohol. Drinking fountains are still provided in many cities and are fed from the municipal supply.

The trend toward fountains that pump and recycle the water in an endless loop is for purely decorative display, although the water is treated to prevent harmful bacteria growth.

A public drinking fountain in the USA, operated via a button at the front.

POWER

A BRIEF HISTORY OF POWER

c.1000 BCE: Simple, single-pulley cranes appear

100 BCE: Waterwheels appear in Greece

15 BCE: The Romans use a vertical wheel for grinding corn

640 CE: The first windmill is powered by a wheel carrying sails

1170: A vertical windmill is developed in Europe to grind corn

1781: James Watt invents the epicyclic gear, which is still used today in many power transfer devices

1792: William Murdock illuminates his Cornish home with gas lights

1807: The first public gas street lighting is used in London

1831: Michael Faraday discovers electromagnetic induction and makes the first electrical generator and transformer

1850: James Francis invents the Francis Turbine

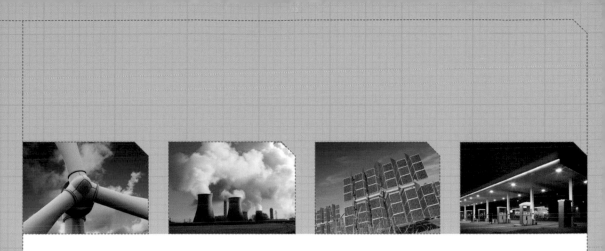

Modern societies rely on a regular and reliable supply of power; without it they would grind to a halt. This chapter describes the many different ways that we harness nature to produce power in a usable and convenient form.

While humankind has produced power on a small scale using water and wind for thousands of years, the landscape of power generation evolved rapidly during the 19th and 20th centuries. Tall chimneys appeared on the skyline as the burning of fossil fuels met our demand for more and more electricity. In recent years, cleaner and more sustainable methods of power generation have become favored, leading to both larger (dams and solar farms) and smaller (residential wind turbines and solar panels) power stations.

As well as illustrating these various types of power station, this chapter will also help you identify the infrastructure of power generation: its extraction, storage, and distribution. Wherever you are in the urban landscape, you will always be able to spot some part of the modern power network.

1880: W. G. Armstrong installs the first domestic electricity supply anywhere in the world

1884: Charles Parsons invents the steam turbine adopted by the world's major power stations

1891: The UK's first AC power system is opened

1933: The UK's national grid starts supplying electricity

1956: The first nuclear reactor becomes operational in the UK

1962: The Telstar communications satellite is powered by solar cells

1977: Solar panels are installed on the White House

1986: A "Level 7" accident at Chernobyl Nuclear Power Plant, Ukraine, spreads contamination across Europe

2011: The world's largest hydroelectric power scheme nears completion at the Three Gorges Dam, China. It will generate 22,500 megawatts

WIND ENGINE

WHAT?
A simple mechanism that harnesses the wind to pump water from wells or drain land of excess water.

WHERE?
There are thought to be 60,000 still in daily use in the USA; they are extremely popular on the farms and ranches in the central plains and southwest of the country. They are also used in other parts of the world: in southern Africa and Australia, farmers still make use of the wind engine.

DIMENSIONS
A modern wind engine, in the right location, will lift 1,600 gallons of water 100 ft (30 m) using a 16-ft (5-m)-diameter sail positioned on top of a tower that is tall enough to capture 15–20 mph (25–30 km/h) wind.

In 1854, David Halladay was asked to design a windmill that could be put to use pumping water. His invention, which required minimal attention, made it possible to raise water from deep underground. Along with barbed wire, this invention helped develop the American West from a scarcely populated landscape into a land of opportunity that could support huge numbers of cattle and provide a higher quality of life for rural Americans.

The Full Works
The idea was taken on, with countless variations and improvements introduced. Eventually, the wind engine was made available as a package, which included pumps, tools, pipes, and farm machinery.

These windmills continue to be sold today and many examples are still in use. The multibladed wind turbine is mounted on a wooden or steel lattice tower and can pump water from wells more than 1200 ft (400 m) deep.

The large number of blades rotate slowly in low winds; this rotary motion is then converted into the up- and downstrokes needed to work the water pumps below. The gearing is designed to be robust and undemanding, requiring an oil change only once

A beautifully maintained wind engine in California.

a year; a clanking, squeaking, grinding noise indicates that this service is overdue.

Taking into account the size of pump needed and the availability of wind at the site, the manufacturer can install the latest in efficient wind pumping equipment which from a distance will look remarkably similar to your neighbor's machine from 50, 60, even 100 years ago.

Installing a Wind Engine
Before a wind engine can be built, a water supply must first be located. This can be done by drilling a hole deep enough into the ground to reach an aquifer—a layer of water-bearing material in which a well can be created. The wind engine supplier will need to know the depth of the well, the level of water within the well, and the height at which the water needs to be accessible. Ideally, the well should be at least 400 ft (120 m) from the nearest large tree or building, and the tower should be 15 ft (4.5 m) taller, to allow maximum wind flow. Wind engine design is not uniform, and elements such as the blades, the tower, the length of pump rod, and the size of the pump will vary.

MUSEUM PIECES

Eight hundred years of windmill construction have left quite a few on the horizon. These large buildings were expensive purpose-built structures that employed state-of-the-art technology to extract maximum power from the wind. Large powerful sails, turned to face the wind, swept round, driving the machinery inside. Many windmills ground corn into flour, while others were used to drain marshland. Today, some windmills remain as monuments to the way we used to live; others, missing their sails, have become unusual homes for eccentric owners, with a grindstone by the front gate.

WIND-DRIVEN WATER SUPPLY

On a windy day, the cistern or an elevated water tower (see picture opposite) can be filled up, creating a pressurized reserve supply for domestic and farm use.

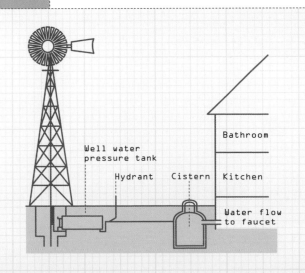

Well water pressure tank

Bathroom

Hydrant   Cistern   Kitchen

Water flow to faucet

WIND TURBINE

WHAT?

A device for turning the power of the wind into the more versatile power of electricity.

WHERE?

On the top of hills, out to sea, or on flat, exposed plains—any place where there is enough wind to make harvesting it worthwhile.

DIMENSIONS

If it's for private use, it's unlikely to be more than 3 ft (1 m) across. If it's for commercial use, such as on a wind farm, it will be an awful lot bigger. Among the world's largest is a prototype in Zaragoza, Spain, which has a tip-to-tip span of 420 ft (128 m), and the Fuhrländer turbine in Laasow, Germany, which is 672 ft (205 m) tall.

Using the wind to grind corn or pump water is all very well and useful, but what humankind really needs these days is electricity. Once we realized that this was possible, windmills, which generate mechanical power, quickly gave way to wind turbines, which generate electrical power. This transformation started in Scotland in 1887, when an academic, James Blyth, found a novel way of lighting his holiday home. The idea took off and wind turbines have been getting bigger and more efficient ever since.

A series of turbines in San Diego County.

Turbine Anatomy

Take one grassy hill with not much around it. Add a number of 200–300-ft (60–90-m)-tall tubular steel towers onto the crest of the hill, then attach three blades 66–130 ft (20–40 m) long to the tower. Allow these blades to rotate at a rate of 20 rpm and there you have your own wind turbine. They are not difficult to spot, even if you paint them gray or white.

Most of the wind turbines you will see are horizontal-axis wind turbines (HAWT). These have the blades pointing into the wind, with the horizontal rotor shaft and electrical generator behind. While small turbines can align themselves using a wind vane, large turbines need a servo motor to turn them into the wind. A gear box turns the slow rotation of the blades into the quicker rotation desirable to drive an electric generator.

THE ÉOLIENNE BOLLÉE

While some hold the view that modern wind turbines despoil the landscape, in the late 19th century a Frenchman named Ernest Sylvain Bollée patented a unique kind of wind turbine that was functional yet undeniably graceful. Known as the "éolienne bollée," only about 260 of this magnificent turbine were built, but if you are lucky enough to spot one (most likely in France, although some were exported) you will not forget it.

Designed for pumping water, a cast iron column containing the pump rods is surrounded by a spiral staircase leading to an ornate iron maintenance platform. This supports the bollée's unique "stator" turbine. The stator is a set of fixed blades that focus the wind onto the rotor blades behind. A small fantail behind the rotor turns the turbine to face the wind. The vast majority were sold to country estates, some of which have been rescued and restored to working order.

Vertical-axis wind turbines also exist in various forms. These have the main rotor shaft arranged vertically, allowing it to be closer to the ground and avoiding the need for it to be pointed into the wind. However, they have a number of disadvantages, including a greater torque stress on the rotor and a lower power coefficient. Look out for "eggbeater" vertical-axis turbines—you can guess what they look like.

Wind Farms

While individual turbines are becoming increasingly popular for single-residence or micro-scale power generation, commercial wind turbines are grouped into two categories: on- and offshore. Wind farms can incorporate hundreds of turbines, all arranged in lines or grids spaced sufficiently far apart so as not to disrupt airflow.

Despite protests by communities who say they don't want wind farms built near them, this renewable energy source is becoming an increasingly familiar sight in countries around the world.

HORIZONTAL WIND TURBINE

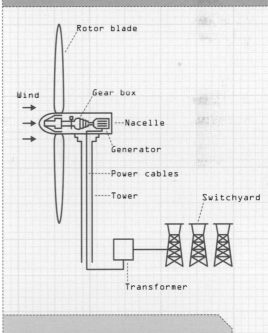

Wind energy is converted to low-speed rotational energy by the rotor blades. A gear box steps the rotation of the blades up to the high speeds required for electricity generation. The generator is housed within the nacelle. This electricity is then distributed to the grid.

PUMPING STATION

WHAT?

Sites that house pumps and equipment for moving fluids from A to B. They often run constantly, moving huge quantities of water or sewage.

WHERE?

Everywhere. They are masters of architectural disguise, ranging from small brick sheds for local drainage to gothic palaces pumping the effluent of the human population.

DIMENSIONS

19th-century steam-driven pumps had huge engine houses, four stories high, to fit the slow, nodding beam engines. Modern pumps are smaller, although they remain substantial installations.

Moving the huge quantities of water or sewage in circulation around a large city is no small task. These operations are carried out morning, noon, and night, during heat waves and snowstorms. Whether the pumping station was built during Abraham Lincoln's presidency or for the 2012 Olympics, it will be a robust building of quality that often reflects the architectural fashions of the day.

The Site Geography

Most of the pipework in a pumping station is buried out of sight. Wherever it is visible, you will see that it is very large and securely constructed. There is often a lot of flat, empty space around an actual pumping station building, but looking down you'll see grids, manholes, and evidence of tanks, pipes, and wet wells. Don't explore a wet well—that's where raw sewage is stored until a sufficient amount has collected to switch the pumps on.

Traditional sewage pumping stations have a wet and dry well, with the pumps installed below the ground. The visible houses contain motors to power the pumps; these have shrunk over the years from huge beam engines to neat electric boxes. The buildings also incorporate the electrical switchgear and other electronics.

More modern pumping stations house the pumps and motors in a sealed unit within the pipes, while the electronics are housed in a small, dull box so as

The pumping station in Nasiriyah, Iraq.

to reduce the visibility of the site. Often, when a site is upgraded with the latest technology, a new pump house is built and the old pumps are kept on standby in case of emergencies. In this way, many pumping stations have a dual significance—practical and historical.

Proven Techniques

An engineer from Boston's Waterworks (pictured below) would recognize the methodology and equipment used in modern pumping stations. Although the materials and technology have advanced and propulsion methods changed, the basic arrangement and process flow have remained remarkably similar. The main power source will differ, of course, with steam used in the 19th century, and electricity used today. The inlet chamber and initial screening traps would be recognizable in old an new alike. The pump chamber, with its wet- and dry-well pumps, would be familiar, although monitoring can today be carried out remotely. Finally, the station pipework, valves, and flow metering, along with the discharge outlet and monitoring, concludes the work of the pumping station. No matter how small modern pumps become, water and sewage cannot be compressed.

LAND DRAINAGE

If the land to be drained is below sea level, the water will have to be forcibly moved using a pump. The Dutch and the British cleared huge areas of wetland in the 19th century, turning it into rich arable farmland. They built magnificent steam-driven pumping stations that today house modern machinery. Elsewhere, a constant battle is fought to remove water that has flooded low-lying areas. New Orleans has a fantastic pumping station; originally built in 1899, it now houses 15 wooden screw pumps that can move over 6 billion gallons (22 billion liters) of water a day.

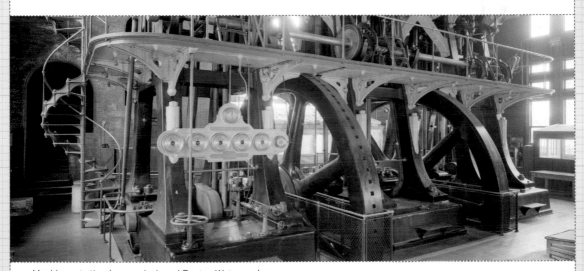

Machinery in the decommissioned Boston Waterworks.

HYDRAULIC POWER

WHAT?

In recent history, water pressure was employed to power cranes, elevators, heavy machinery, and anything else that required lifting. Clean and silent, one remote power plant could provide power to lots of applications.

WHERE?

London and Geneva had networks supplying power to all kinds of businesses. Large dockyards were the main users, and swing bridges often had small hydraulic power plants. One such plant in Lincolnshire, UK, is now a private house.

DIMENSIONS

The London Hydraulic Power Company's pipe network is 150 miles (240 km) long and is buried just below the streets. The pipes, 2–10 in (5–25 cm) in diameter, run through a subway near Tower Bridge and extend from the East End docks to Kensington in the West End.

Before the advent of electricity, a hydraulic power network supplied the pressure to operate machinery without having to also run a power plant. The power was delivered on demand and paid for according to usage, just like electricity today. Used to power mainly lifting devices in thousands of hotels, stores, offices, docks, and factories, it could also run water motors, which in turn operated a vast array of machinery.

Powering the Docks

The port of Grimsby in the UK was the first major dock to use hydraulic power and the 309-ft (94-m) tall tower used to create the necessary water pressure can still be seen. From 1852, it supplied power to machinery for the lock gates, dry docks, and to 15 quayside cranes. These large towers were replaced by the accumulator, a simple device consisting of an extremely heavy weight above a cylinder full of water, which, via a hydraulic line, creates a constant pressure.

Pumping stations and accumulators were exported to docks across the globe to power dockside machinery, fire hydrants, and to raise and lower bridges. Hydraulic power was utilized by thousands of businesses until it was replaced with electricity.

Most, if not all, hydraulic power networks have now been decommissioned, although it's likely that the pipes remain in place. Numerous small buildings that were once part of city-wide networks are still standing, along with the power stations. In Wapping, London, one of the original six hydraulic power stations has been converted into a restaurant and arts center. Known as the Wapping Project, the center exhibits art among the original machinery.

The Jet d'Eau in Geneva, now one of the world's largest fountains, was originally built in 1886 as a safety valve for the Geneva hydraulic power network. The water jet, which can reach a height of 98 ft (30 m), has became a popular landmark and was retained despite the closure of the network.

POWER

FOR CLEANING CARPETS

The hydraulic power station in Wapping, London, provided power to many local businesses during its lifetime. The central turbine hall (pictured here) produced the energy for stores, theaters, hotels, and private homes. The high-pressure water was also used for other purposes; for example, it was used to create the vacuum in a latter-day vacuum cleaner. The operation of the cleaner was virtually silent and the dirt was disposed of directly into the drains. By using the hydraulic main purely as a power line to create a vacuum, alongside water from the mains, a high-pressure fire-fighting system could be created. A large hydraulic injector fire hydrant could deliver 650 gallons (2,460 liters) a minute via hoses or a sprinkler system protecting warehouses and public buildings.

HYDRAULIC ACCUMULATOR

The hydraulic accumulator is the water-powered equivalent of an electric battery. It works by exerting pressure on the piston and fluid by means of a raised weight. When power is needed, the accumulator delivers the pressurized fluid more smoothly than a pump. It can deliver huge force instantaneously, which is ideal for raising and lowering heavy loads. A small pump can be used at times of low demand to recharge the pressurized system.

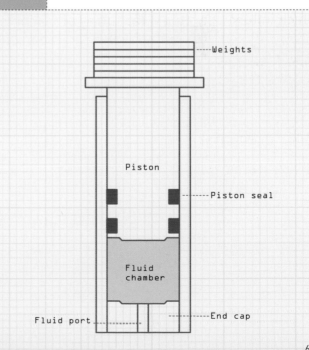

Weights

Piston

Piston seal

Fluid chamber

Fluid port

End cap

GEOTHERMAL POWER

WHAT?
Tapping Earth's molten core to heat your bath water or power your hair dryer.

WHERE?
Wherever cracks in the Earth's crust bring this source of heat close enough to the surface to be exploited safely.

DIMENSIONS
They can stretch more than 2 miles (3 km) downward into the Earth's crust. A geothermal development in California, USA—known as The Geysers—spans an area of around 30 square miles (78 km²) and incorporates 22 geothermal power plants, drawing steam from more than 350 wells.

Our planet is an abundant source of heat. Aside from keeping us warm, heat energy can also be used to create steam, which can then be used to generate electricity. While we have been exploiting this geothermal power for thousands of years, it has only recently been harnessed to generate electricity.

The West Ford geothermal power plant in San Francisco.

Spotting natural geothermal power is not difficult in some parts of the world—volcanoes, geysers, and bubbling mud pools are all telltale clues. But spotting it being harnessed in an urban or industrial setting isn't so easy.

Developing Natural Power
Italy produced the early pioneers of geothermal power. In 1904, Piero Ginori Conti created a geothermal generator that produced enough electricity to power four lightbulbs. This small breakthrough quickly led to the world's first commercial geothermal power plant being built in 1911, in the Italian town of Larderello. Italy remained the only country to operate such power plants until 1958, when New Zealand commissioned the Wairakei power station. This was closely followed by a plant in California in 1960.

GEOTHERMAL POWER FOR HEATING

Electricity generation is not the only use we have for geothermal heat. If it is readily available, this clean, sustainable, and relatively cheap source of energy can be used to heat entire towns using coordinated district heating schemes. Ninety-five percent of all housing in Iceland is heated in this way. Not wanting to waste any of that heat, the spent warm water is then piped below sidewalks to melt snow. Geothermally heated water can also be used in swimming pools. Although the UK is not known for its volcanic activity, hot springs beneath the city of Bath have been used to provide warm outdoor bathing all year round since Roman times. Industry also benefits from this source of natural heat, particularly those in which atmospheric temperature control plays an important role, such as with large commercial greenhouses.

Each of these early power plants was built in areas of high geothermal activity, where groundwater aquifers combine with geothermally heated rocks to produce naturally occurring hot water. As this water is brought to the surface by wells drilled into the aquifer, it loses pressure and turns into steam. Unsurprisingly, geothermal power is most commonly utilized for electricity generation in countries with a lot of geothermal activity. Iceland and the Philippines are alone in producing more than a quarter of their total electricity from geothermal power.

Identifying Geothermal Plants

Large bore pipes snaking along above ground and the presence of steam are sure signs of a geothermal power plant. The pipes direct the steam from the aquifer to the central generating plant. White vapor clouds are produced whenever excess steam is vented, and many plants are notable for their large cooling towers.

If you have hot rocks below the surface, but no natural aquifers to produce steam, you can pump water underground to produce it. Two bore holes are drilled, one to pump the water in, and the other to extract the steam. To generate electricity effectively, the steam from water should be no cooler than 300°F (150°C); if the geothermal heat underground is insufficient to produce steam at this temperature, then other measures are necessary. A heat exchanger can be used, and a fluid with a lower boiling point than water can be pumped onto the rocks instead. In a neat twist on the concept of renewable energy, the steam released as waste by the generator can be used by a secondary generator to produce yet more energy.

As technology improves, geothermal energy is being sourced from greater depths, although this isn't without risk: drilling in Basel, Switzerland, was stopped in 2006 after it triggered earthquakes.

SOLAR POWER

WHAT?
Converting the sun's energy into electricity, which can then power your refrigerator or satellite.

WHERE?
Photovoltaic cells have been added to calculators, house roofs, and office blocks, to name but a few.

DIMENSIONS
An individual solar cell can be as small as a few millimeters across. Joined together, they form a solar panel (or module) about 3 ft (1 m) across, which can be combined into a solar "array" many feet across. Solar farms, consisting of many arrays, cover hundreds of acres.

The sun is nature's power station. It drives the wind, the waves, ocean currents, and sustains life with light and warmth. Surely some of this power can be utilized to satisfy our ever-increasing need for electricity? Creating electrical power from the sun seems like an obvious way to go, but we are still struggling to achieve this cheaply and efficiently enough to revolutionize power generation.

The Photovoltaic Effect
Most solar-powered electricity generation is based on the photovoltaic (PV) effect—the creation of an electric current in a material through exposure to light. The first functional solar cell was constructed by an American, Charles Fritts, in the 1880s. A precursor to the ubiquitous silicon solar cell was produced in 1954, with an efficiency of about 5 percent. Fifty years of development has brought that efficiency up to around 20 percent for a typical commercial PV panel cell.

A growing concern about our impact on the environment, coupled with financial incentives from national governments, have made domestic PV panels increasingly common and thus easy to spot. Most look the same—rectangular frames with a flat, blue network of cells pointing south. The most efficient alignment and angle is dictated by the hemisphere, the season, and the latitude. You can also spot small panels powering utilities along roads and railroads, where power isn't readily available. These are often coupled with small turbines to provide electricity when there is wind rather than sunshine.

Not all solar power is generated through photovoltaic cells. Concentrated solar power (CSP) systems use lenses or mirrors to focus a large area of sunlight into a small beam, which then heats water and drives a steam turbine to produce electricity. Many track the sun as it moves across the sky.

Solar Farms
If there is enough space, the sun's power can be farmed. This consists of huge numbers of either

photovoltaic solar panels or CSP systems. The largest photovoltaic solar farm is in Ontario, Canada, although it provides relatively little power, with a capacity of 65 megawatts (MW). CSP farms are more productive and therefore more common. The largest is in the Mojave Desert in California, USA, with a capacity currently of 354 MW. Spain is home to other large CSP farms.

All of these large CSP farms use parabolic troughs. A parabolic trough is a thermal energy collector consisting of a long parabolic mirror with a tube running along its length. Sunlight is reflected by the mirror and concentrated on the tube, thereby heating a thermal transfer fluid held inside the tube. The trough is usually aligned on a north–south axis, and rotated to track the sun as it moves across the sky each day. The heat transfer fluid is superheated and then directed to a steam turbine to produce electricity.

Solar Heating Panels

Solar panels are also used to heat water, usually on a domestic scale. As solar heating panels carry liquid, they tend to be thicker. In hot countries, you will often see a solar panel with a cylindrical tank mounted on roofs. These "close-coupled solar hot water" systems are easy to install and use gravity to circulate the water.

BUILDING-INTEGRATED PHOTOVOLTAICS

Research is being carried out on the possibility of integrating photovoltaic cells into construction materials. Not only does this save time and money, it also reduces the visibility of the solar technology. A good example is the roofing tile: rather than adding solar panels to a roof once it has been built, tiles have been developed that are themselves made from photovoltaic material. In fact, entire facades are now being made in this way. Glass is being developed that harvests the sun's energy as it passes through. This has been made possible by the development of thin film solar cells, which are light, durable, and flexible, and therefore far more versatile than previous materials.

DOMESTIC SOLAR POWER

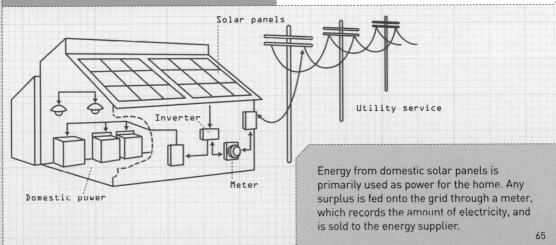

Solar panels

Utility service

Inverter

Meter

Domestic power

Energy from domestic solar panels is primarily used as power for the home. Any surplus is fed onto the grid through a meter, which records the amount of electricity, and is sold to the energy supplier.

TIDAL POWER

WHAT?

Weird and wonderful apparatus that attempts to harness the power of the tides to produce electricity.

WHERE?

Any stretch of water where the tide has an extensive range and produces strong currents.

DIMENSIONS

Most of the hardware is underwater. A tidal barrage is clearly visible, the largest of which is off the Brittany Coast in France and is 2,500 ft (750 m) long.

Sources of renewable energy are not the most reliable: even in the middle of a desert, the sun doesn't shine all the time, and on the top of a hill it isn't always windy. Tidal power, however, is different. Twice a day, every day, a huge volume of water rushes toward the coast, raising the level of the sea, before rushing out again. Tidal power is a reliable energy source, but harnessing it is an ongoing challenge that has produced a number of competing techniques.

You won't see tidal power generators along a stretch of coastline. The most common location is in an estuary or between islands, where seawater is compressed as it is channeled by the two landmasses. This increases the flow of water and range of the tides, concentrating the available energy.

Types of Tidal Power

There are two types of tidal power generator: tidal stream generators and tidal barrages. Tidal stream generators are basically underwater wind turbines that utilize the flow of water rather than the wind. As water is denser than air, these "tidal turbines" have the potential to generate far more energy per unit than wind turbines. A number of different designs are being developed and deployed, but all you are likely to see of them are the tops, which have warning lights to alert boats and ships to their presence.

Tidal barrages are far easier to spot. They act like a one-way dam, allowing water to flow freely through in one direction, then using sluice gates to channel the return flow through turbines to generate electricity. Very few tidal barrages have been built, although a number of others have been proposed. The largest is currently the La Rance plant off northern France. Barrages raise greater environmental concerns than tidal turbines because of the disruption to flora and fauna it causes within the enclosed estuary.

Don't be fooled by the Thames flood barrier at the mouth of the River Thames in London. This may look like a tidal barrage, but its purpose is to stop particularly high tides from flooding central London. It doesn't generate electricity and has a far less restrictive effect on water flows than a tidal barrage.

TIDAL MILLS

Although tidal power has not been a particularly effective way of harnessing natural energy, we have been using it since the Middle Ages. Tide mills are a variation of the water mill, but instead of the millpond being filled by a river bypass (or "leat") it was filled by the tide. They were usually built in the shelter of estuaries with a high tidal range. A section of an inlet was closed off with a dam to form a pool. A one-way gate let the seawater in as the tide advanced, then shut itself when the tide receded. Next to the pool was the mill and water wheel. When the difference in water level between the sea and the pool was great enough, the sluice gate was opened and the miller started milling, even if it was the middle of the night. Tidal mills had their heyday in the Middle Ages in Ireland, the UK, France, and Portugal, although some can be spotted along the Atlantic coastline of North America.

TIDAL BARRAGE

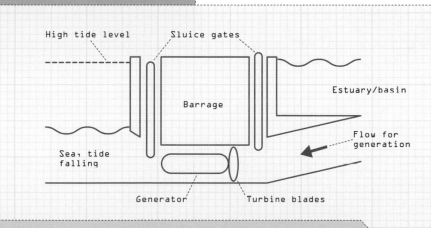

Tidal barrages make use of the tides to convert energy into power. After a high tide, water is held behind a dam in an estuary. As the tide recedes, the water drives turbines, which in turn drive generators. The system is essentially the same as a river dam, but uses the tides rather than rainfall to replenish the water supply.

HYDROELECTRIC POWER

WHAT?
Often huge structures that channel water to produce a rapid, high-pressure flow, which can then be exploited to produce electricity.

WHERE?
On mountains and in rivers—any place where large volumes of water can be accessed and controlled.

DIMENSIONS
Ranging from "micro-hydro" units that power a house or small community to huge hydroelectric dams that power entire nations.

There is a massive amount of energy stored in the world's fresh water. This energy is being utilized across the world to produce electricity. From a small generator dropped into a stream to power a single dwelling to the massive engineering project that is the Three Gorges Dam in China, hydroelectric technology provided 16 percent of the world's power in 2009 without producing any waste.

While water has been used to power machines for millennia, it wasn't until the invention of the electric generator in the late 19th century that it was used to produce electricity. The Industrial Revolution brought rapid advances in hydroelectric generation; the Schoelkopf Power Station No. 1 near Niagara Falls in the USA started an exponential rise, with over 200 additional power plants appearing in the USA in the following eight years. Today, countries including Norway, Congo, Brazil, and Paraguay generate over 85 percent of their electricity using hydropower.

Hydroelectric Generation
All hydroelectric power plants use the combination of a water turbine and generator to turn the potential energy of water into electricity. The greater the potential energy, the more electricity; therefore, power plants are positioned in places where the potential energy is greatest. This is usually within a dam, so if you've spotted a dam, look for telltale power lines leading from it.

"Run-of-the-river" hydroelectric stations are set within a river that has no associated water storage area. Most are smaller than dammed plants, but they are usually in large rivers with little seasonal variation in flow. The lack of dam and reservoir means that there is little or no upstream flooding of land, which reduces the environmental impact and helps to keep local residents onside.

Underground hydropower stations such as the Churchill Falls plant in Canada do away with the river altogether. Instead they make use of high mountain lakes, diverting the river that would normally flow from the lake down a long underground pipe to the turbines and generator before discharging it downstream. If you see a waterfall that looks like it used to have an awful lot more water in it, there may be an underground hydropower station nearby.

THE LARGEST POWER STATIONS IN THE WORLD

Large hydroelectric power stations are not just physically very big, they also produce an awful lot of electricity. In terms of power generating capacity, seven out of the top ten power stations in the world are hydroelectric.

Rank	Station	Country	Capacity (MW)	Fuel type
1	Three Gorges Dam	China	18,200	Hydroelectricity
2	Itaipu Dam	Brazil/Paraguay	14,000	Hydroelectricity
3	Guri Dam	Venezuela	10,200	Hydroelectricity
4	Tucuruí Dam	Brazil	8,370	Hydroelectricity
5	Kashiwazaki-Kariwa Nuclear Power Plant	Japan	8,212	Nuclear
6	Bruce Nuclear Generating Station	Canada	7,276	Nuclear
7	Grand Coulee Dam	USA	6,809	Hydroelectricity
8	Longtan Dam	China	6,426	Hydroelectricity
9	Krasnoyarsk Dam	Russia	6,000	Hydroelectricity
10	Zaporizhzhia Nuclear Power Plant	Ukraine	6,000	Nuclear

MICRO-HYDROPOWER

Micro-hydropower generation occurs all over the world, often in remote communities or developing countries. Their appearance varies depending on location and available materials, but they work just like their big brothers. All will include a head of water from a pool, stream, or river, a pipe called a "penstock" to take the water from the high point to the low point, a turbine to turn the flow of the water into mechanical energy, and a generator to produce electricity for nearby homes or communities.

OIL PLATFORMS—OFFSHORE

WHAT?
An oil rig or platform is a structure whose purpose is to drill wells, extract and process oil and natural gas, and bring the products to shore. In a marine environment, this isn't easy.

WHERE?
Oil exploration and drilling is taking place across the globe on a vast scale. In 2006, there were 3,858 oil and gas platforms in the Gulf of Mexico alone, and there are over 100 active rigs out in the hostile environment of the North Sea.

DIMENSIONS
As we run dry the wells close to shore, the platforms have to get bigger. At 2,001 ft (609.9 m), the Petronius oil platform in the Gulf of Mexico is one of the tallest free-standing structures in the world. Only 250 ft (75 m) of the structure is actually above water, however.

Our dependence on fossil fuels has led us to seek them out in difficult locations, where extracting them is dangerous, expensive, and tests our engineering ingenuity to the maximum. The solutions vary from location to location, depending on the depth of water and the size of the field.

Oil Exploration and Extraction
Mobile drilling rigs are used to establish whether there is sufficient petroleum in a particular stretch of the seabed to justify building a more permanent structure. Of these, the jackup is the most common kind. Typically costing $1.8 billion, it is transported out to the drill site, where it then lowers three or four massive legs to the seabed until the rig body lifts clear of the water. Barges are also used in this exploratory stage.

Offshore production platforms are brought in once it has been established to an acceptable degree of certainty that enough oil is there. Designed to last decades, they are partially constructed near land before being towed into position. Under the sea the rig may be supported on fixed legs or moored with guy lines in position. Above the water will be a functional steel platform supporting a latticework tower, a helicopter landing pad, cranes and storage vessels, workshops, living accommodation, and often a flare burning off waste gas.

Flotels
Oil rigs are cities on the sea, self-sufficient in energy and water, running electrical generators and water desalination plants. In order for the rig to operate 24 hours a day, several hundred people are employed in 12-hour shifts, seven days a week, and a great deal of effort has to go into making life tolerable. In order to keep your skilled workforce happy and working hard, private rooms, satellite TV, gyms, and saunas are often available, along with food around the clock. If you book into a flotel, you'll find yourself on a floating accommodation block moored to an oilrig, so be prepared to work very hard.

POWER

OIL PLATFORMS

Conventional fixed-platform and compliant tower rigs are supported by piles driven into the seabed; the deeper the water, the greater the forces on the support structure, which is designed to be flexible, bending but not breaking under great loads. Greater depths require the platform to float on huge ballast tanks, towed into position then anchored by chain or wire rope mooring lines, vertical in the case of the tension-leg platform. By altering the amount in the tanks, the platform can be maintained in a stable float position for drilling.

Floating production systems do not actually drill but are used for extended periods in these inhospitable environments for processing and storage.

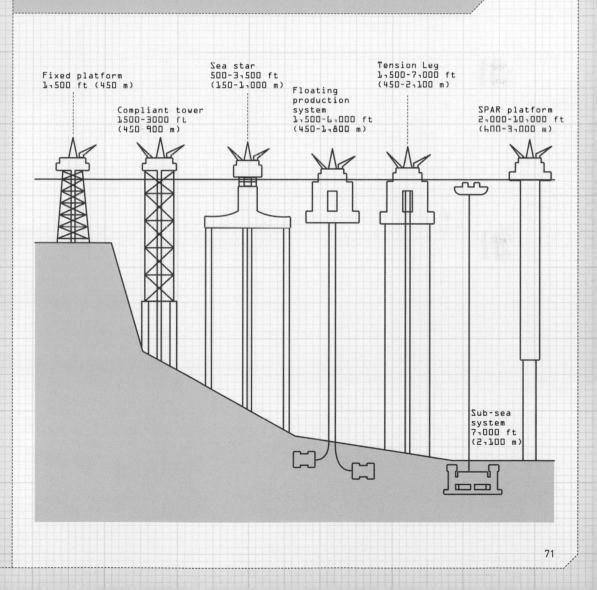

Fixed platform
1,500 ft (450 m)

Compliant tower
1500-3000 ft
(450 900 m)

Sea star
500-3,500 ft
(150-1,000 m)

Floating
production
system
1,500-6,000 ft
(450-1,800 m)

Tension Leg
1,500-7,000 ft
(450-2,100 m)

SPAR platform
2,000-10,000 ft
(600-3,000 m)

Sub-sea
system
7,000 ft
(2,100 m)

PUMPJACK

WHAT?
Pumpjacks, also known as Thirsty Birds, Nodding Donkeys, and Horse Heads, have lots of nicknames, but they all refer to pumping oil using a counterbalanced beam.

WHERE?
Rusting away on abandoned fields, grazing the surface in great numbers where there's oil below. There are hundreds on the Lost Hills Oilfield near Route 46 in the USA, and one in the middle of the high rise buildings of Salvador, Brazil.

DIMENSIONS
Built to last, their robust metal beams are a crude version of 19th-century beam engines. Size varies according to the depth and volume of the oil reserve, and the rate of extraction required.

The fine detailing of the pumpjack varies between manufacturers, but the see-saw beam giving the nodding donkey appearance is characteristic of all. The pump rods are attached at one end and a counterbalance and driving force at the other. The design has hardly changed since Walter W. Trout filed his patent in 1926.

Workhorse
Used inshore for relatively low-yield oilfields, the pumpjack is reliable and low-maintenance. The low-volume "stripper" wells in which pumpjacks are found can draw a maximum of 15 barrels a day, while the average is just 2.2 barrels. Nevertheless, these wells make up 84 percent of US domestic oil wells and together produce 20 percent of all domestic oil.

The nodding action of the pumpjack draws the oil to the surface on the upstroke. Today's donkeys are generally electrically driven, although in remote areas diesel still provides the motive force. On the surface, oil is separated from any gas or water before being despatched to the refinery.

The life of a pumpjack, operating outdoors come rain or shine, is a harsh one that they stand up to remarkably well. Despite their functional design, they present a striking sight when silhouetted against the sunset. Occasionally they are redecorated, being transformed into insects with antennae, thirsty toucans sipping drinks, and rodeo-riding characters.

A trio of pumpjacks in northeast China.

POWER

THE OIL PATCH WARRIORS

In the middle of an area of ancient woodland in Nottinghamshire, England, stands the UK's first commercial oilfield. Discovered in 1939, by the end of World War II some 170 pumpjacks were in place—one to every two and a half acres. They produced an average of 64 barrels a day. The field was eventually pumped dry and closed in 1989. The original woodland has regrown; it is now a nature reserve with rare orchids, oak trees, and five nodding donkeys left as a memorial to American oilmen who worked there during the war.

THE PUMPJACK

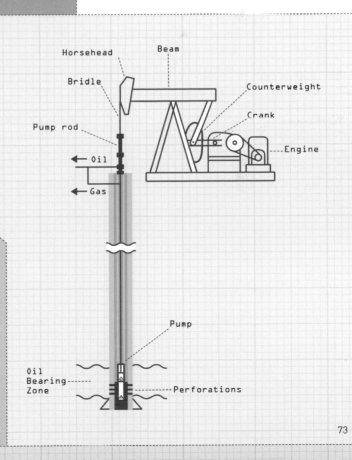

The rotary motion of the engine drives the crank and counterweight in a circular motion, which moves the tailend of the beam up and down. This drives the horsehead, which in turn drives the pump rod down through a shaft. At the base of the pump shaft, perforations in the cement casing allow the oil to be sucked into the shaft and up to the surface.

OIL REFINERY

WHAT?

An oil refinery takes the raw material of crude oil and creates gasoline, as well as lots of other fuel products.

WHERE?

Ideally far from residential areas, with lots of energy available and ready access for the transport of materials. Many refineries are near seaports or large rivers to facilitate transport and provide the large amount of water needed for the oil-refining process.

DIMENSIONS

Oil refineries operate on a huge scale; even the smallest process a hundred thousand barrels of crude oil a day. The largest refineries currently operating are in Iran, Venezuela, and South Korea.

A refinery works around the clock, operated from a highly automated control room and employing hundreds of people. A barrel of crude oil has all kinds of useful things in it that have to be separated out in order. These go on to make different types of fuel oil as well as providing the basic ingredient of most plastics. The refinery tackles everything: splitting, processing, storing, and dealing with the waste.

Separation by chemical or fractional distillation is carried out in the distinctive tall columns that dominate the refinery. The sprawling industrial process disguises very little and the network of pipes moves the product around for further treatment. Conversion involves cracking or rearranging the molecules to create products depending on customer demand; for example, turning diesel fuel into gasoline. The quantities are controlled remotely and the process is highly automated. Treatment plants remove impurities and prepare the waste products for disposal.

The Refinery Layout

A large area is generally given over for the storage of the end products before final distribution by pipeline or shipping, or, in the case of smaller quantities, barge, road tanker, or rail.

Gas flares act as a safety valve, burning off excess gas. Easily visible at night, a small amount of gas may be burned continuously, like a pilot light, so that the flare is constantly ready should the gas-processing equipment become over pressurized. Steam can be injected into the flame, which reduces highly visible black smoke but increases the sound of the volatiles burning to a loud roar. Cooling towers are another highly visible feature, dominating the skyline with their sheer bulk contrasting with the wisps of steam.

The seemingly chaotic combination of pipes, tanks, chimneys, and gas flares may not create the most visually appealing landmark on the horizon, but it is without doubt a modern engineering masterpiece, and it produces the fuel which literally helps human civilization to operate.

HOW BIG IS AN OIL REFINERY?

The quick answer is very big, and also very expensive, as the Dung Quat Refinery in Vietnam illustrates. Phase one of construction took over three years and was completed in 2009, at an estimated cost of $3 billion. The refinery's processing, utility, and off-site facilities cover over 270 acres (110 hectares); storage tanks and the gas flare-off area occupy 192 acres (78 hectares); and the interconnecting roads and pipelines cover 98 acres (40 hectares).

The refinery has a total capacity of 7 million tons (6.5 million tonnes) of low-sulfur crude, but this still only provides 30 percent of the country's domestic requirement.

FRACTIONAL DISTILLATION

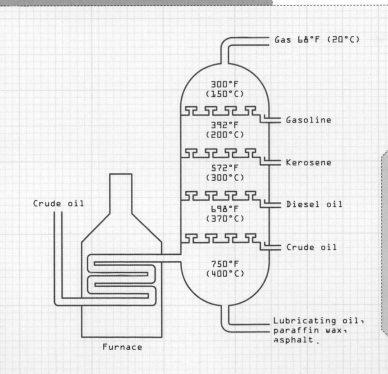

Gas 68°F (20°C)

300°F (150°C)

Gasoline

392°F (200°C)

Kerosene

572°F (300°C)

Diesel oil

698°F (370°C)

Crude oil

750°F (400°C)

Crude oil

Lubricating oil, paraffin wax, asphalt.

Furnace

The most important part of the refining process is the separation of the crude oil into its constituent substances. This is achieved by heating the crude mix to a temperature of 750°F (400°C). These substances can be processed further to produce a wide variety of fossil-fuel by-products.

GAS STATION

WHAT?

A place to fill your vehicle up with fuel. It is known variously as a gas station, gasbar, petrol bunk, service station, filling station, and garage.

WHERE?

Everywhere with motor vehicles; there are 40,000 in Japan alone.

DIMENSIONS

The number of pumps at a gas station varies; typically, there would be around five or six, spaced evenly across a forecourt 50 ft (15 m) wide.

For our cars and trucks to travel from A to B they need fuel. Originally all gas stations offered full service, with attendants to clean your windshield, check your tires, and wish you a good day. This kind of service is becoming more and more rare; it is actually forbidden in New Jersey and Oregon. Modern filling stations may ask you to pay before you pump, or to select the right fuel type from a bewildering choice. Some involve no human contact at all, just a machine that takes your money and pumps the required amount of fuel.

The first drive-in filling station was opened in Pittsburgh in 1913, and since then most filling stations have been built to a similar design. Most of the storage tanks are hidden underground; pump machines are located so your vehicle can drive alongside; and a small building contains a payment kiosk and sells both essential items and temptations for your journey. A canopy protects you from the rain, and tall signs that are easy for the driver to spot at a distance advertise the brand and often the price of fuel, along with other services.

Gas Station Architecture

Most gas stations are uninspiring, functional places, where no one lingers longer than the time it takes to pump and pay, but there are quirky, beautiful, and even breathtaking variations all over the world.

In 1937, Danish architect Arne Jacobsen designed a beautiful modernist gas station that is still functioning in Copenhagen today. In Los Angeles, Frank Gehry split opinion with a geometric canopy that replaced a run-down old structure. Over in Beverly Hills, the Jack Colker Union 96 gas station has a sweeping compound-curve canopy originally designed for LAX airport.

In countries with a long history of automobile use, the number of filling stations is declining, in part thanks to the improved fuel efficiency of modern cars. The original "Model T" Ford, released in the USA in 1908, could travel only 15 miles (24 km) to the gallon. This modest distance necessitated the presence of gas stations in every town.

POWER

GOOGIE GAS STATIONS

"Googie" is an architectural style that drew heavily on the 1950s and '60s "car culture." With its bold lines and geometric shapes, it seemed to anticipate an exciting, technologically advanced future for humankind. The humble gas stations offered the perfect blank canvas for experimentation, with sweeping cantilevered roofs, bold angles, and plate glass helping to capture this bold vision of the future. The aesthetic form didn't obstruct the functioning of the station and, importantly, drew the attention of passing drivers. Ultimately, this forward-looking style became dated, and many examples have been torn down or redeveloped by big brands, who have stamped an "identikit" corporate image across the land.

MODERN GAS STATION LAYOUT

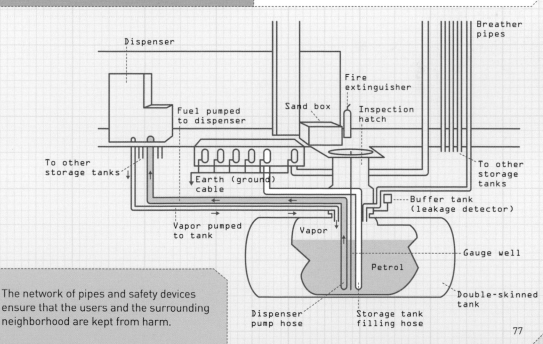

The network of pipes and safety devices ensure that the users and the surrounding neighborhood are kept from harm.

PIPELINE

WHAT?

A hollow tube through which anything chemically stable, whether gas, liquid, or slurry, can flow or be persuaded to move through the application of pressure.

WHERE?

Across the globe. Fossil fuels have to be transported from the inconvenient, inaccessible places where they are found to processing plants, and on to cities where they are used.

DIMENSIONS

Anything from 2–60 in (51–1,500 mm) in diameter, stretching across cities, countries, and even continents.

Fossil fuels are getting ever more expensive. It is therefore imperative to establish a cheap and reliable means of transporting them. Pipelines provide this, even in often hostile environments.

A Secure and Efficient Method

A pipeline buried underground offers security because it is protected from extreme weather conditions and the risk of sabotage. The capital cost for laying the pipe may be substantial, but it will still be cheaper than constructing a road or river along which to transport oil, tank by tank. Oil is pumped through steel or plastic tubes at 3–20 ft (1–6 m) per second, with pump stations positioned along the route to boost the flow rate. Shut-off valves allow sections of pipe to be isolated for maintenance. Leak detection systems are regularly employed to protect

TECHNICAL TERMINOLOGY

Whether it's fossil fuels, water, or beer, the main elements of a pipeline assembly are the same.

Initial injection station	Where the product enters the system. Surrounded by storage containers and pumps or compressors.
Compressor or pump stations	If it's a liquid there will be a pump, while gas pipelines require a compressor. These move the product along the pipeline. Their location is determined by the nature of the product and its destination.
Block valve station	These are simple safety devices that allow a section of pipe to be isolated for repair.
Regulator station	A special type of safety valve through which pressure can be released. It is often located on the downward slope after a rise.
Final delivery station	The terminal from where the product can continue its journey to the customer via a tanker or a distribution pipe network.

the environment and prevent costly losses. By law, in the state of Washington, a leak of 8 percent of the maximum flow must be found within 15 minutes. Given the quantities and distances involved, this is no easy matter; a combination of computer modeling, satellite surveillance, and physically walking the line is employed.

A single pipeline can accommodate a succession of different products. There is some mixing at the "interface," but given the volumes, this is easily absorbed. The Calnev Pipeline, which stretches 550 miles (890 km) across the USA, carries gasoline, jet fuel, and diesel from Los Angeles all the way to Las Vegas. Coal can also be transported this way. By mixing it with water to create "slurry," it is possible to pump coal considerable distances. The fuel is dried it out at its destination. A 325-mile (525-km) slurry pipeline is currently being constructed to transfer iron ore from the Minas-Rio mine in Brazil to a port for shipping.

A PIG IN THE PIPELINE

A pipeline offers a low-tech solution with a long life span, but every now and again some maintenance has to be undertaken. The solution is to send in a Pipeline Inspection Gauge or PIG (shown above in a section of pipe). Setting off from "PIG-launcher" stations, they travel along the pipe, scraping clear material that has built up over time. Smart PIGs are the brains of the bunch, inspecting and recording the condition of the pipe to identify metal fatigue, corrosion, and other damage.

NORTH SEA PIPELINES

The Forties pipeline system branches out under the North Sea, joining over 50 locations to two hubs, which pump oil through a 36-in (90-cm) pipe. The 105 miles (169 km) of underwater pipeline carry 700 thousand barrels of oil ashore a day—nearly half of the UK's daily oil requirement.

North Sea

TOWN GASWORKS

WHAT?
Gas plants in our cities.

WHERE?
Originally on the outskirts of urban areas, often close to a river and close to other industry. As our work patterns change, industrial areas are often redeveloped into housing and retail neighborhoods. The gasworks site is often the last to be redeveloped.

DIMENSIONS
These were large industrial factories and storage areas covering large areas of urban landscape. The largest in Europe was on the Greenwich Peninsula, London, which grew to over 240 acres (97 hectares).

Before natural gas became the major power source we know today, gas was manufactured directly from coal and tar, and processed and stored in each major town and city. Operating for up to a hundred years, the gasworks produced toxic sludge that contaminated the land, leading to expensive clean up operations prior to redevelopment.

Old Pollutant

The old town gasworks were substantial sites that incorporated train lines and river access, huge processing plants, and storage facilities, along with support services and offices. The soil contamination caused by the sites means that they are often the last areas in a city to be redeveloped. The underground storage tanks and soil often contain a toxic legacy that has to be removed and cleaned.

Poor record-keeping and weak legislation relating to waste disposal mean that it is often not known what contaminants may be residing underneath derelict gasworks site. Cleaning contaminated land can be extremely expensive and few developers are willing to take on the risk. The cost involved, and the technical challenges that the job entails, means that gasworks are often only redeveloped as part of a larger-scale regeneration scheme by the civil authority.

The Greenwich Peninsula in East London was the site of an extensive gasworks that operated between 1889 and 1985. In 1996, with funding provided by central government, the slow process of redevelopment began with a small-scale exhibition. Very slowly, retail parks, housing, and transport infrastructure have been built, and the area is providing facilities for the 2012 Olympics.

All major cities and towns face the problem of contaminated land, and the solutions vary from place to place. If a large chunk of prime real estate is on the market at a very cheap price, it is advisable to look very carefully at what might be underground.

GASWORKS PARK

In Seattle, a 20-acre (8-hectare) gasworks site has been converted into a public park that preserves some of the machinery and buildings as monuments. A brave decision at the time, the magnificent views of downtown Seattle include major elements of the manufacturing plant. The park has retained other elements of the sites. Six steel natural gas generator towers remain, along with their processing towers, an oil absorber, and an oil cooler. At 20–80 ft (6–25 m), the generator towers cast a distinctive silhouette on the landscape. The wooden pump and boiler house have been retained, along with some of the plant, and a covered picnic area has been constructed on the site. It has become a unique landmark in Seattle.

The gasworks in Luleå, Sweden.

GASHOLDER

WHAT?

Engineers call them gasholders; you and I know them as gasometers. It is a simple structure that allows for the safe storage of gas at a controlled pressure.

WHERE?

Most commonly used in the UK and Germany. They are a distinctive structure on the urban skyline, often situated near the center of cities. The famous column-guided gasometers have witnessed every cricket match at the Oval, London since 1853.

DIMENSIONS

The Oval gasometer is 200 ft (60 m) across and holds up to 1.75 million cubic yards (50,000 m³). The tallest gasometer in Europe towers 384 ft (177 m) above Oberhausen in Germany and has been reinvented as an exhibition center.

There are two main types of gasholder: those that you can see moving down during peak demand, and those that remain rigid. The rigid, waterless type was an early design that after 200 years is being revisited with oil, grease, or membrane seals to contain the gas.

The telescoping type is the more familiar design, of which there are two distinct categories: column-guided, or guide-framed, which move up and down within a rigid framework; and spiral-guided, which has no frame and is guided up by concentric runners inside the previous lift. There is also the telescoping cylinder type, which is essentially a gas-filled "balloon" that rises and falls within a seal of water.

Safety Concerns

Early concerns over the safety of gasholders led to strengthened buildings, gasometer houses, being built around them. The logic behind this was not particularly sound since small leaks of gas became trapped between the gasholder and the gasometer house, resulting in a potentially explosive combination of gas and air trapped in a small space. Never popular in the UK, in Europe—where extreme cold weather is more of a risk—huge ornate gasometer houses were constructed like the beautiful survivors in Vienna and Berlin.

In 1927, a workman carrying out routine maintenance on a Pittsburgh gasholder, the largest in the world at the time, inexplicably used a blow torch near to a leak, thus blowing the whole thing into the sky. The sound was heard 20 miles (32 km) away and the initial explosion produced a cloud of burning gas 100 ft (30 m) across, which rose above the city before burning out. The surrounding buildings were destroyed and 28 people were killed, with hundreds injured by flying glass.

The UK is now in the process of dismantling many of its urban gasholders, although some have become such a prominent part of the skyline that they are being retained as historic landmarks.

CONVERTING GASHOLDERS

As energy companies turn to alternative storage methods, gasometers are being dismantled all over the world, allowing vast areas of land to be redeveloped, often in prime locations. However, many gasometers have become an important part of the urban landscape and inventive ways have been found to give them a new lease of life.

Four giant gasometer houses in Vienna have been converted into offices, shops, and apartments. The King's Cross telescoping frames in London are due to be reinvented as an "architectural space" and the gasholder in Oberhausen, Germany, is now an exhibition hall. It enjoyed a brief period as a concert hall, thanks to its excellent acoustics.

GASOMETERS

Gasometer house

Gas

Water

Line from production

Distribution

Gas

Water

Line from production

Distribution

Cased inside a fixed gasometer house, the sealed piston rises and falls according to how much gas is being held. Large amounts of gas can be stored for times of peak use and the gas is available at a standard pressurized supply via the delivery network.

BIOMASS POWER

WHAT?
Grow it, burn it, and make some electricity. A power station running on fuel that was, until relatively recently, still growing.

WHERE?
Close to a very large source of organic fuel, be it forest, sugar cane fields, or garbage.

DIMENSIONS
Biomass power plants tend to be much smaller than their unsustainable fossil fuel relations.

We all know that fossil fuels are running out, which is a shame because they're rather useful. Those trees just keep on growing, so why don't we burn them? In fact, there are all sorts of things that were recently growing that can be used for fuel. While biomass crops are becoming more popular (or notorious) as a fuel for vehicles, they are also being used for electricity generation, albeit on a relatively small scale at present.

The Difficulties With Biomass

Biomass power is becoming popular because it is renewable (we keep producing waste and plants keep on growing), the fuel is relatively cheap, and the environmental and financial cost of disposing of waste in other ways is increasing. Some argue that it isn't as environmentally friendly as it sounds because the fuel is still being burned, which produces greenhouse gases. Because the fuel has a relatively low thermal rating (i.e., you have to burn an awful lot of it to achieve the necessary temperature), transport costs can be relatively high compared to fossil fuel plants.

Because of the transport issue, the largest biomass power plants are in countries with large areas of managed forest. Finland hosts two of the biggest: Alholmens Kraft and Kaukaan Voima power stations. In common with most other large biomass plants, they use a combination of virgin wood, wood waste, paper pulping waste, and other organic waste to generate electricity and heat for district heating schemes (this is known as combined heat and power, or CHP). Biomass plants appear much like other thermal power stations—you should be able to spot a fuel storage area, fuel processing, the plant itself, chimneys, and electrical transmission apparatus, although the power lines may run under the ground. On site, there will also be a fuel processing plant to grind the organic matter into a powder prior to burning. There may be conveyors for transporting the fuel about the plant.

The location of a biomass plant will be largely dependent on the type of fuel it uses: those that burn garbage will be in urban areas; those that burn wood (and products of wood) will be near forests and paper pulp mills; and those that use sugar cane waste (called "bagasse") or other crop residues will be in arable growing areas.

In addition to their use in biomass power stations, certain crops, once treated, can also be used as fuel for vehicles. Known as biofuels, they fall into two main categories: bioethanol is produced by fermentation of the sugar or starch in a plant and is more commonly used in North and South America; biodiesel is made from oils extracted from plants such as soya and is more common in Europe. Both fuels are usually blended with conventional gasoline. Biofuel-processing plants are characterized by a large vertical reactor and holding tanks. They can vary in size from small "farmyard" processors to large commercial factories.

Whether a small-scale project involving a single farmer or a large-scale process utilizing forests countrywide, biomass fuel generation is one of several "old" technologies being reinvented for a world seeking alternatives to fossil fuels. The rising cost of energy has made research into alternatives necessary and presents an opportunity to turn previously unheralded resources into valuable assets.

FROM MUCK TO MEGAWATTS

Enterprising farmers are using waste from livestock and dairy waste to generate power and heat for themselves and their neighbors, and selling any excess to the power grid.

This form of biomass power relies on anaerobic digestion, where microorganisms break down biodegradable material in the absence of oxygen to produce biogas—a mix of methane and CO_2. The biogas is then burned to produce electricity and heat. To produce the gas, slurry and other liquid waste is pumped into a large cylindrical digester. The biogas is piped off to a storage tank and on to the burner and generator. The remaining matter is spread on the fields as fertilizer.

Thousands of farms across the world are now generating power in this way; if you see an abnormally large gas cylinder among the cows, pigs, or chickens, you may have spotted a biogas power plant.

BIOMASS POWER PLANT

The fuel is stored and processed prior to delivery to the boiler. Flue gases are filtered to remove particulate matter prior to exhausting through the flue. Electricity is generated by a steam turbine and distributed to the grid.

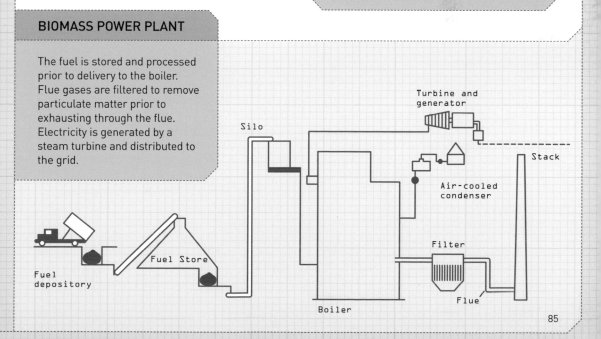

Fuel depository

Fuel Store

Silo

Turbine and generator

Stack

Air-cooled condenser

Filter

Boiler

Flue

FOSSIL FUEL POWER STATION

WHAT?
A big industrial complex where coal, oil, or gas goes in and electricity, steam, and smoke come out.

WHERE?
Not attractive or clean places, fossil fuel power stations are best placed away from people.

DIMENSIONS
The generating plant, fuel storage, chimneys, and cooling towers together cover a large area.

Unfortunately, the kind of electricity we need to power our everyday needs such as TVs, coffee grinders, and waffle irons doesn't grow on trees or fall into our laps. It has to be generated, and we are now so dependent on electricity that it has to be generated in huge quantities. For many years, the most convenient and efficient way to do this has been by burning fossil fuels—coal, oil, and gas.

The Process

The process of electricity generation is well established. The fuel is burned to heat water to produce steam under high pressure. The steam is then condensed using cooling water, which turns a turbine, which spins a generator to produce electricity. The condensed water is returned to the boiler to start the process over, while the temperature of the coolant is lowered via a cooling tower or nearby lake or river, or it is put to use to provide district heating. Since the first commercial power plant, the Edison Electric Light Station, was opened in London in 1882, scientists and engineers have been working on making this process progressively more automated, more efficient, and, in recent years, cleaner.

At the start of the process is the fuel. Be it coal, oil, or gas, it has to be transported to the plant. Coal is the easiest to spot as it is often in huge piles adjacent

to the plant and is delivered by road, rail, or barge. Coal trains can be up to 1.25 miles (2 km) long with 10,000-ton (9,000-tonne) capacity, which it delivers onto conveyor belts that feed the plant boilers. Oil is delivered by tanker or pipeline and stored in vertical hopper tanks. Gas is delivered by large bore pipeline.

Next is the power plant itself. Not much to spot there except a very big square industrial building. More eye-catching are the large chimneys and cooling towers nearby that stretch upward. Flue gases from the boiler are vented via one or more of these chimneys, which are often painted in red and white stripes to increase visibility. These emit hot, translucent smoke, with most of the black particulate matter having been removed earlier by a precipitator. These hot gases rise quicker than the white steam billowing out from the cooling towers. "Wet condensing" cooling towers, with the familiar curved sides and open mouths, use the natural draft effect.

Smaller box-like structures are likely to be forced-draft cooling towers.

Finally, the electricity has to be taken away via a transmission network—a mass of carefully routed power lines carried away across the landscape by stately pylons.

Decommissioned Plants

Fossil fuel power stations can stagger on into old age; many in operation today have been in service for 50 years or more. The oldest are often mothballed and only fired up for testing or during periods of emergency demand, while newer, cleaner, and more efficient power stations bear the brunt of demand. It is often strict national carbon emissions controls that finally kill off these dinosaurs. Similarly, due to associated health concerns, power stations are being built further from population centres. Large urban power stations such as the iconic Battersea Power Station in central London are largely a thing of the past.

COOLING TOWER ART

Cooling towers are not hard to spot. They rise up to 650 ft (200 m) from the ground and emit huge clouds of steam. While most of them are a dull gray concrete color, you may be lucky enough to spot one that has had a facelift from a forward-looking official.

Cooling tower art was born in World War II, when the British government commissioned artists to camouflage power stations to make them less visible to German bombers. Since then, cooling towers have used as advertising space, and they have been painted to appear more "eco-friendly" or simply to look more attractive.

COAL-FIRED POWER STATION

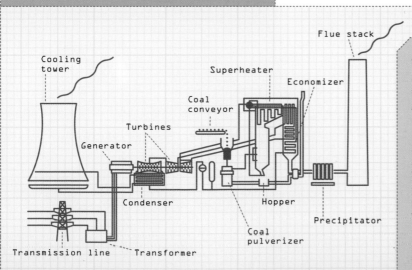

The coal is pulverized in a hopper before being fed into the furnace. Above the furnace, water is superheated in the boiler and the resulting steam is passed through a series of turbines. The electricity passes through a transformer to the transmission lines. Waste gases are filtered by a precipitator to remove dust before venting through the stack.

ELECTRICAL POWER TRANSMISSION AND DISTRIBUTION

WHAT?
Wires that take electricity from the power station to homes and businesses.

WHERE?
All over the planet, even under the sea. In these electricity-dependent times, it is difficult to escape the network of power lines spinning their web above our heads.

DIMENSIONS
Pylons stretch up into the sky carrying large cables well out of reach. By the time they reach our homes, they have shrunk considerably, hidden within armored subterranean lines.

It's a long way from the power station to the electrical socket in your wall. All that electricity has to be transported from the generators to homes, factories, railroads, and streetlamps without leaking, killing anyone, or getting in the way. To make the job more difficult, different users have different needs—AC or DC, high voltage, low voltage, single-phase, or three-phase. All of these problems were addressed in the early 20th century with the advent of the grid transmission and distribution system.

As electricity consumption increased in the late 19th century, power lines started to appear above towns and cities. In those days, spotting a power line in a rapidly expanding city like New York would not have been difficult—lighting, motors, and heavy machinery all had different voltage requirements and so had to be supplied from separate generators. This produced a web of power lines on cluttered poles.

Serbian inventor Nikola Tesla found a solution—the transformer—that allowed the stepping up and down of alternating currents to different voltages. This invention drove the development of our regional and national power transmission and distribution systems. Electric power transmission is the bulk transfer of electricity from generating power plant to substation. Electrical distribution is the transfer from substation to point of use.

Electrical Power Transmission
Once it has been generated, electricity is difficult to store, hence the need for power grids. These are networks of interconnected transmission lines that allow the smoothing of demand over very large areas and allow you to still make a cup of tea or coffee if part of the grid, including a generating plant, goes down. These grids are very big. While the UK and Ireland have their own, most of continental Europe shares a single grid.

Transmission is via high-voltage three-phase power lines. High voltage means low current, which means less energy lost along the way. The two most visible parts of the transmission network are pylons and substations. Transmission substations look like cages of several large buzzing metal insects with bodies of transformers and metal frames, legs of wires and insulators. Don't venture too close, as they have a deadly sting.

Electrical Distribution

The distribution grid takes over when substations step voltages down to 50,000 volts. At this stage, smaller power plants, particularly renewable generators that generate lower wattages, can still contribute power, but the main purpose is distribution to the end users. Although substations have the same components as the transmission system, they are smaller and more numerous, and poles are used in place of pylons. As they reach the end of their journey, power lines disappear underground. In urban areas, lines are often buried as they leave the distribution substation; in rural areas, it may continue along a pole, dropping down underground just outside the property.

When spotting parts of the transmission and distribution network look out for a triangular yellow warning sign showing a bolt of lightning; it is used internationally to denote high voltage.

ELECTRICITY DISTRIBUTION

Transmission Grid

The transmission grid is a network of interconnected power-generating plants of capacity 100 MW and above. These include nuclear, oil, coal, and large hydroelectric power stations. They are joined via high-voltage transmission lines and substations. Factories with especially high electricity demands may be directly connected to the transmission grid.

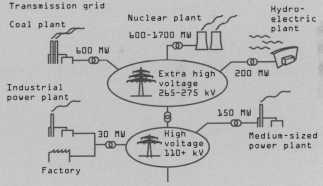

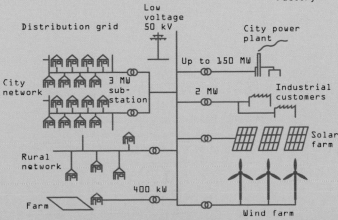

Distribution Grid

This is the lower-voltage part of the system that connects the transmission grid to end users such as factories and homes. Lower voltages result in greater energy loss, but they are safer and so used in urban areas. Smaller-scale power generating plants are connected directly to the distribution system.

TRANSMISSION TOWER

WHAT?

Steel lattice towers marching across the landscape carrying power lines.

WHERE?

Everywhere. Supporting power lines run from generating plants to substations, for onward distribution to our homes and factories.

DIMENSIONS

The size, height, and distribution of towers varies in response to the landscape and the load they are expected to carry, allowing endless permutations of their form. They usually range between 49 and 180 ft (15 and 55 m) tall, although look out for the superstructures, which can be over 980 ft (300 m) tall.

Considered visual pollution by some and elegant towers of engineering endeavor by others, transmission towers are an essential part of the electrical supply grid. To do their job, they have to keep high-voltage power lines well out of reach of people and well separated. They must hold the lines aloft whatever the weather may throw at them, across mountains, rivers, and plains.

Although transmission towers can be constructed from wood and concrete, steel offers the greatest variety and flexibility. Steel lattice towers can be assembled on site, allowing extremely tall structures to be built. Steel monopole towers are durable and easily manufactured, but they have to be transported preassembled, which limits their height. Monopoles are increasingly used in urban areas.

Any tower you observe can be classified by the way in which they support the power lines:

Suspension towers support the lines vertically through suspension insulators.

Tension towers are like suspension towers but support the lines horizontally through tension insulators.

Terminal towers are used for the transition from overhead to underground power lines, or for the connection of lines to a substation. As the strain on such towers is one-sided, they need to be anchored to the ground.

Transposition towers are only used for long stretches of power line where the lines have to cross over to maintain equal lengths between phases.

Lattice towers are often assembled on site. A pulley system, or "jin-pole," is attached to the tower as it rises into the air, lifting the pieces up for assembly. Monopole towers are usually preassembled from sections and hoisted up to vertical in one complete piece. Preassembled towers can also be transported by helicopter in particularly inaccessible areas.

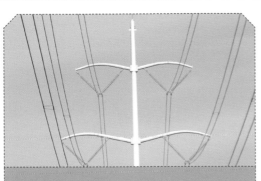

ELECTRICAL FIELDS

Overhead power lines generate significant electric fields that extend down to ground level. Invisible to the eye, the field can be detected using a fluorescent light tube. The field created by the 400,000-volt cables will excite the atoms of mercury gas in the tube, making it emit ultraviolet light. This invisible light strikes the phosphor coating on the glass tube, making it glow. People, weather, and the angle at which the tube is held will affect the light's intensity. If you jump in the air you'll even experience a small shock.

Types of Current

You can tell whether the tower you have spotted is carrying AC or DC voltage through its power lines. AC power lines will be high-voltage and three-phase; therefore the towers will be carrying lines in multiples of three. DC power lines will usually be in pairs, or even a single line. To keep you on your toes, if it is by a railroad, it might be single-phase AC and so have pairs of lines. Most towers will have an identification plate or other means of allowing the electricity company to know which of the hundreds in their care they are looking at.

In response to complaints that the towers are an eyesore, architectural engineers across the world have created some beautiful and elegant designs that can give the engineering-minded spotter unexpected pleasure. Don't assume that all pylons are alike or you will miss out.

These transmission towers, stretching as far the eye can see, have an almost otherworldly beauty.

ELECTRICAL INSULATOR

WHAT?

Small glass or ceramic components that prevent the electrical current traveling from the wires onto the poles and transmission towers that support them.

WHERE?

Originally on the telegraph system, they can now be seen on transmission towers, utility poles, and any other utility that carries an electrical charge.

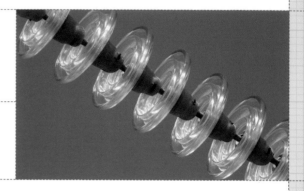

DIMENSIONS

From a few inches in length to several feet on large electrical installations.

Electrical power transmission cables suspended in the air are bare, except where they enter buildings; at points where they are supported by utility poles or transmission towers, insulators are required. Electricity will seek out the path of least resistance and without insulators the metal support towers would be dangerous, live electrical conductors.

Born of Necessity

The insulators used for high-voltage power lines must be durable, waterproof, and resistant to sunlight and extremes of temperature. Glass and ceramic have been used successfully for over 150 years. The characteristic smooth glaze on porcelain conductors acts as a seal that prevents the conductor from soaking up moisture that would lower its resistance. Many companies that supply industrial high-voltage ceramic insulators have diversified from the manufacture of other earthenware goods such as plates and bowls. Insulators were first used when the only electricity being carried across the land was to enable signals to be sent via telegraph. The damp wooden utility poles directly supporting the electrical wire caused the signal to degrade substantially—a problem that was solved by the introduction of insulators.

Over time, insulators will naturally accumulate particles of dirt and salt, as well as collecting rainwater. This decreases their effectiveness. To help prevent this, the shape of insulators on high-voltage lines is corrugated so that water runs off more quickly.

High-voltage transmission lines use a modular system called "cap and pin." By arranging the ceramic caps vertically in a series, the insulator can be easily

A cap-and-pin electrical insulator.

adjusted according to the voltage—it is simply a case of increasing or decreasing the number of caps. As the voltage increases, so must the number of caps. For example, if you see an insulator with eight caps, the line voltage will most likely be 155 kV; if the insulator has 23 caps, the line will be carrying 360 kV.

Maintenance

Whenever a problem arises with a high-voltage power line, the task of fixing it isn't without difficulty. To avoid any kind of contact with the ground, an engineer will be lowered from a specially equipped helicopter to perform repairs while suspended in the air. In this way, work can be carried out on the power lines, allowing the engineer to physically inspect and repair any damage while minimizing the risk of electrocution. In the case of the insulators, routine cleaning has to be carried out to remove the mess left behind by birds perched on the wires and ceramics. In the event of a power surge, a buildup of potentially conductive matter would increase the possibility of electricity "jumping" across the insulator (see box, right). The cleaning operation is carried out using a helicopter rigged with a specialized high-pressure water jet.

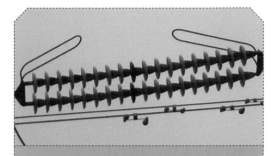

ARCING HORNS

In the event of a power surge, electricity will jump, or arc, across the insulator, destroying it and causing the high-voltage lines to fall to the ground. To prevent this, two prongs, known as "arcing horns," are positioned at either end of insulator. Under normal conditions, the gap between the horns is too wide for the electricity to bridge. In the event of a surge, however, they allow the electricity to bypass the insulator, thereby protecting the line from significant damage.

HIGH-VOLTAGE SAFETY DEVICES

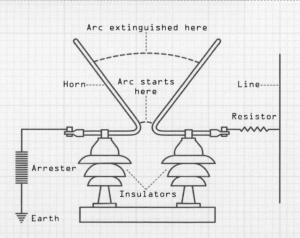

Arcing horns are protective devices and always come in pairs. Constructed from conductive material, they project up from either side of the insulator. The conductivity of the material is gauged carefully so as to provide an escape route for electricity only under extreme conditions.

NUCLEAR POWER

WHAT?
A power plant that generates electricity using energy produced by nuclear fission.

WHERE?
Well away from populated areas. Often next to the coast in order to use seawater as coolant.

DIMENSIONS
They are relatively small when compared to non-nuclear power stations of the same generating capacity.

One of the more emotive sources of power, to many people nuclear power will forever be linked with bombs, meltdown, fallout, and other doom-laden images. However, nuclear fission is an extremely efficient way of generating power and it produces negligible carbon emissions. It seems the world is divided into nations that have it, nations that want it, and nations that want nothing to do with it.

The Development of Nuclear

Despite the trepidation and mystery surrounding the use of uranium to generate heat, nuclear power stations work in a similar way to many other power plants. A fuel is used to heat water into steam, which drives steam turbines, which turn generators, which produce electricity. Simple. The problem is that nuclear fission can be highly unstable if it is not tightly controlled and if something does go wrong the consequences are potentially catastrophic.

After research into nuclear fission before and during World War II, the 1950s saw commercial nuclear power plants commissioned in Russia (then the USSR), the UK, and the USA. Capacity grew quickly during the 1960s and '70s, then slowed during the 1980s. This was partly in response to falling oil prices but also due to increasing opposition to nuclear power after serious incidents at Three Mile Island, USA, in 1979 and Chernobyl, Ukraine, in 1986.

The nuclear reactor is the core of the power plant. Uranium fuel rods generate heat through the fission chain reaction. This heat is transferred to carbon dioxide gas or high-pressure water being pumped through the reactor. A heat exchanger then generates steam to be fed into the turbine. The rate of reaction is controlled by boron "control rods," which absorb neutrons. When the rods are lowered into the reactor, they absorb more neutrons and the fission process slows down. To generate more power, the rods are raised, increasing the rate of nuclear fission.

Common to many other power stations, nuclear power stations are often most easily identified by the large cooling towers emitting clouds of white steam. Seawater can also be used as a coolant, removing the need for the towers. Instead, look for large numbers of power lines and pylons fanning out from the plant. Reactors often have domed roofs. If in doubt, just look for the familiar three-cone radioactive sign.

POWER

NUCLEAR POWER

Heat is generated within the reactor vessel. The rate of reaction, and therefore the heat generated, is controlled by the control rods. A closed loop of pressurized water is pumped through the vessel and on to a steam generator.

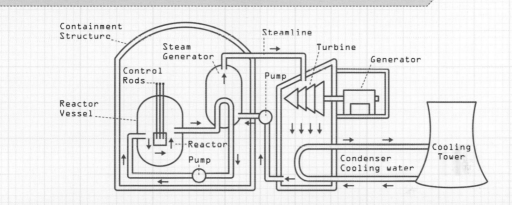

COUNTRIES WITH NUCLEAR POWER PLANTS

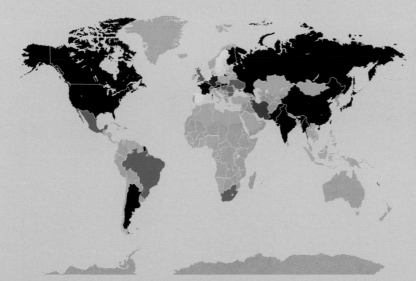

You simply won't be able to spot a nuclear power plant in many parts of the world. Several countries have decided not to pursue the nuclear option (Austria has made them illegal), others would like to but don't have the technology, and an increasing number are in the process of building them.

■ Operating reactors, building new reactors
■ Operating reactors, planning new build
■ No reactors, building new reactors
  No reactors, planning new build
  Operating reactors, stable
■ Operating reactors, considering phaseout
■ Civil nuclear power is illegal
  No reactors

DISTRICT HEATING

WHAT?
A system of heating that distributes hot water from centralized boilers.

WHERE?
Anywhere, although it is particularly popular in areas that have naturally occurring sources of heat, such as Iceland.

DIMENSIONS
From a few solar panels connected to a row of houses, to a power station heating an entire city via pipes 3 ft (1 m) wide.

We are used to having mains water, gas, and electricity, so why not mains heat? District heating networks provide heat in much the same way as central heating: through pipes in the ground and available at the turn of a valve. That may sound odd to some, but to the majority of people in Denmark and Iceland it is entirely normal. The drive to reduce emissions of CO_2 has helped to reawaken interest in the district heating scheme, and its popularity is increasing.

Most modern power generation is all about producing electricity using steam. This process, which requires very high temperatures, produces large quantities of hot water as a by-product. Rather than release it into the atmosphere via a cooling tower, the water can be redirected along insulated pipes to provide heating for nearby homes. The sensible option of using waste heat from power generation to run a district heating scheme is called combined heat and power (CHP).

This system is used around the world in coal, oil, nuclear, and geothermal plants, so if you spot a power station there is a chance that it operates according to CHP. As CHP is considered "greener" than standard power generation, it is often advertised. This is especially true of incinerators, which can make themselves more popular by combining waste disposal with electricity and heat generation.

This waste incinerator in Munich, Germany, operates a CHP scheme.

POWER

Once you have spotted the power plant, look for the hot water or steam distribution pipes. While there are many different pipes to spot in an urban environment, these will be heavily insulated to minimize heat loss. Pipe diameters step down as they get closer to the end user, starting at 3 ft (1 m) wide and narrowing to 1 in (25 mm). Just like electricity distribution networks, the district heating scheme employs substations to regulate supply. Finally, some kind of metering system, usually at the point of use, will allow those being kept warm to be charged.

Not all district heating schemes are from CHP plants. Smaller-scale centralized heating schemes, supplying a few houses or streets, can be carried out by biomass boilers, solar panels, or anything that heats up water.

What happens during the summer when no one wants the heat? The system can be routed through refrigeration units to provide district cooling. District cooling systems can also use cold water from the sea or deep lakes to cool buildings.

Heat usage can be measured by a "heat meter"; however, it is much simpler and cheaper to use a water meter to record the amount of water passing through a building.

NEW YORK STEAM

Have you ever wondered why steam can be seen rising from the streets of New York in movies and TV shows? If so, you've already spotted a district heating system. The New York Steam Company began providing steam to buildings in lower Manhattan in 1882. Today, it supplies steam for heating, hot water, and air conditioning to buildings across Manhattan, including the Empire State Building. As steam is supplied, not hot water, it can also be used for cleaning and disinfection. The pipes carrying the steam run under the streets. If water comes into contact with the hot pipes or valve gears on a cold day it evaporates, sending a cloud of steam up through the nearest manhole.

DISTRICT HEATING

Hot water or steam is supplied to the customer via a feed pipe and returned in a closed loop to the generating station via a cooler return pipe. The system can be switched over to refrigeration during hot weather.

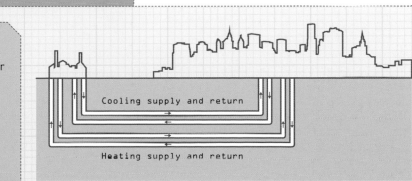

Cooling supply and return

Heating supply and return

TRANSPORT

A BRIEF HISTORY OF TRANSPORTATION

40,000 BCE Basic dugout canoes

5000 BCE First use of sails in Ancient Egypt

4000 BCE Stoned paved streets in Ur, Middle East

2000 BCE First use of Kongming lantern, prototype of the hot-air balloon in China

600 BCE Wagonways (horse-pulled wagons on tracks) in Greece

312 BCE Straight stone roads of the Roman Empire

700 CE Tar-paved roads in Iraq

1088 Magnetic-needle compass for navigation first described in China

1760 First use of metal railroad tracks, north England

1770s Steam technology applied to boats

1783 First human flight in hot-air balloon in France

1812 First commercially successful steam locomotive in England

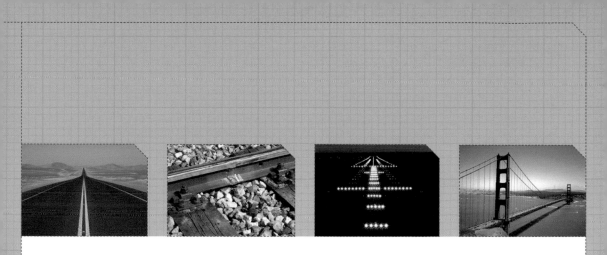

One of the aims of this book is to familiarize you with what you may see from the window of any train, plane, or automobile. This chapter focuses on the infrastructure that supports these channels of circulation—the great networks that reach all the way around the world.

The simple urge to reach out beyond one's immediate surroundings has been one characteristic of man's post-cave development. Perhaps the most profound invention of them all—the wheel—has enabled us to travel from A to B for millennia, and improvements in technologies mean X, Y, and Z are now equally accessible destinations. Travel has broadened the collective mind, enabled empires to flourish (and fall), brought people together, and driven them apart. The routes developed have created a network of ways, paths, tracks, and lanes that criss-cross the air, the ground, and the water—a truly worldwide web. The ingenuity and creativity employed in our bid to go further applies equally to the big and the small: the design of the highway is just as interesting as the suspension bridge it crosses, for example. This chapter gives due consideration to all.

1869 Transcontinental railroad completed in the USA

1869 Suez Canal completed

1901 Patent granted for tarmac

1902 First controlled, powered, sustained flight by the Wright brothers

1912 *Titanic* sinks off Newfoundland

1914 Panama Canal opened

1964 Opening of the first high-speed rail link between Tokyo and Osaka, Japan

1969 First flight of Concorde, the world's first supersonic passenger airliner

2003 Last commercial flight of Concorde

2004 The world's tallest vehicular bridge, the Millau Viaduct, opens in France

ROADWAY

WHAT?
A network of constructed paved routes that support vehicular transport.

WHERE?
The greatest density of roadways is found in industrialized countries, but paved roads are found all around the globe.

DIMENSIONS
From 10 ft to 320 ft (3 m to 100m) wide, with a total length estimated to be well in excess of 15 million miles (25 million km).

The world's roadways are probably the most obvious and least observed elements of the transport system. Hardly any of us can leave our homes without stepping or driving on paved roads, and yet we are largely unaware of their history or how they are constructed. Here we will look at these general aspects of roads, and in subsequent pages we'll examine the special cases of highways and freeways.

Modern roads are built to withstand climatic extremes.

Even though permanent roads were first constructed—of stone paving or timber—more than 5,000 years ago, and the Romans developed a network of stone paved roadways that enabled them to control a vast empire, most transport traveled on dirt roads until well into the 19th century. In summer even the best coaching roads would deteriorate into potholed trenches full of dust, while the winter turned them into bogs. This was largely true even in the towns, although the use of rounded stones, or cobbles, from the 15th century on improved the situation in some of the larger towns and cities.

Road Structure
In the late 18th century, the blind Scottish road builder John Metcalf developed a system that used a base of large stones covered with earth and stone and topped with gravel. He built more than 180 miles (290 km) of roads in the north of England, and these proved to be well drained and hard-wearing. In

France, the engineer Pierre-Marie-Jérôme Trésaguet developed an equally effective two-layer system using large stones topped with smaller stones. In the early 1800s, these methods were refined by two Scottish engineers, Thomas Telford and John McAdam. Telford's great contribution was to raise the road up and give it a curved center, or camber, while McAdam invented a smooth surface of crushed stone and gravel that allowed the water to run off. This construction method, known as "macadamizaton," spread rapidly throughout Europe, and in the USA it was first used on the National Road from Maryland to Illinois, completed in 1838.

The Birth of Blacktop

The great breakthrough in road surfacing came in the 1830s with the introduction of pitch and tar to bind and stabilize the dusty top layer of the "macadam" road, but the idea didn't take hold until the beginning of the 20th century, when the growing use of bicycles and then automobiles led to a huge demand for more and better roads. In 1901,

Edgar Hooley patented a surfacing method using a combination of tar, pitch, resin, and cement mixed with aggregate (sand, gravel, and crushed rock) that was then compacted using a steam roller. He called the material Tarmacadam, or Tarmac, but this term is now often used to refer to the material that developed from it—asphalt concrete, or blacktop, the world's most commonly used road surface (see p. 28).

Concrete Roads

The second most important road surfacing material is concrete. Although this material has several disadvantages when compared with asphalt concrete—being more expensive to build and repair, and providing less grip in wet or snowy conditions—concrete roads last up to four times as long without maintenance, give better vehicle fuel consumption (because they deflect less and absorb less energy), and are less susceptible to damage from spills of oil or fuel, or from extremes of temperature. Concrete is also considered to be less environmentally damaging than asphalt, which is derived from fossil fuels.

THE PLANK ROAD

One of the earliest known roads—the 5,800-year-old Sweet Track across a marsh in southwest England—was made of timber, but wooden roads have been used in much more recent times. Corduroy roads, trackways of whole logs laid across the path, were built by loggers in western Canada so that horses could haul the timber from the forests over soft ground, and several important 19th-century transport routes in the northeast and midwest of the USA were made of sawn planks. In parts of Australia, too, wooden planking (usually hard-wearing eucalyptus wood) proved to be a cheap way to build roads over rough terrain, and some were still in use in 1940.

SIDEWALK

WHAT?
A walkway beside and distinct from a road, dedicated to pedestrian use only.

WHERE?
Alongside lanes, roads, streets, avenues, and boulevards in every village, town, and city.

DIMENSIONS
Often in proportion to the roadway they accompany, sidewalks range in width from the barely passable to broad thoroughfares that accommodate outdoor seating for cafés and restaurants.

The idea of separating the traffic on the roadway from the pedestrians walking at the sides certainly goes as far back as the Romans, and there are still 2,000-year-old examples of roads with a raised sidewalk on each side. However, after the fall of the Roman Empire their engineering achievements fell into decline, and as urban populations grew and horse-drawn transport increased in the cities of Europe, conditions on the streets deteriorated.

Poorly drained roadways, inadequate sewage systems, and a great many horses made town and city streets a nightmare to walk on during the 17th and 18th centuries, especially in the winter. Gradually, throughout the 19th century, paved sidewalks were built in the cities to enable pedestrians to keep clear of the mud and filth and to avoid the danger posed by horses and carriages traveling on roads that were still made of dirt and gravel. Indeed, in Britain the term "pavement" refers specifically to the sidewalk and not to the roadway.

In the smaller towns throughout Canada, Australia, and the USA, many business owners took matters into their own hands and invited passers-by out of the mud and onto raised wooden sidewalks along the storefronts, often protected from the weather by an overhanging roof or jutting upper floor of the building. Now seen only in Western movies, many

wooden sidewalks were still in use into the 1950s, and examples have been preserved in places such as Old Town Sacramento, California, and Skagway, Alaska.

With the improvement of roads and the building of many new ones as towns and cities expanded over the course of the 19th century, sidewalks became

Sacramento's Big Four House, with its covered sidewalk.

TACTILE PAVING

In many countries, especially those with effective disability/accessibility legislation, certain pedestrian hazards are indicated on the sidewalk in ways that can be detected by people with visual impairment. Known by such terms as tactile indicators and detectable warnings, these consist of raised markings—small domes, flat-topped cones, or raised strips—that can be felt with a stick or even the soles of the feet, and they are built into the sidewalk at locations such as busy road crossings or changes of level. The idea was first implemented in Japan, in 1967, and in many Japanese cities tactile trails will lead you for miles along the streets, with marked turnoffs to public buildings, and all the way to the spot on the station platform where the train doors will open for you.

an essential element in urban planning and came to fulfill a number of roles. The basic design differs little around the world—sidewalks are generally surfaced with slabs, concrete, or asphalt, are raised slightly above the level of the roadway, and are edged with an upright curb. The curb, usually made of stone or concrete, demarcates the edge for pedestrians, creates a barrier to vehicles, helps to form a drainage channel at the side of the roadway, and prevents water from penetrating the sidewalk.

The building of sidewalks also created a new kind of social space, especially in the expanding cities, where there was room to build impressive avenues with wide pedestrian areas on each side, taking their cue from such magnificent European boulevards as the Champs Élysées in Paris. Dating from the 17th century, this avenue was the first to be paved with asphalt (back in 1824) and its sidewalks are 30 ft (9 m) wide.

Café and restaurant owners have incorporated the sidewalk into their seating plans, storekeepers place their wares there to attract the buying public, street vendors make it their place of business, and musicians and artists come to entertain. The addition of trees and grass verges provides welcome shade and green space in many urban areas, and the sidewalk has become the location for a host of services and amenities above and below the ground, as evidenced by the mailboxes and manholes, streetlights and fire hydrants that dot its length.

Spacious sidewalks are part of the character of the Champs Élysées.

BIKE LANE

WHAT?

Paths and lanes created expressly for use by cyclists (sometimes shared with pedestrians) and from which motorized traffic is generally excluded.

WHERE?

In the world's larger cities and towns, but especially in North America, Europe, Australia, New Zealand, and South Africa.

DIMENSIONS

Just a few feet wide, but getting longer by the day.

Cyclists can take much of the credit for the rapid improvements in road infrastructure that took place in Europe and North America toward the end of the 19th century. As the bicycle became an increasingly important means of transport, the cycling lobby made stronger and stronger demands for a smoother ride, and gradually they got what they wanted—but then they had to share it with the automobile.

Nineteenth-century bicyclists had their differences with travelers by foot and by horse, but the advent of motorized road users brought far greater conflict, each claiming that the other threatened their safety and ruined their enjoyment of the thoroughfare. In Britain the peddlers called for the building of an entirely separate "motor road" system, while in Germany the drivers wanted to ban bicycles from the public roads. "Segregated bicycle facilities" provide some of the solutions to the problem of shared road use.

On-Road Cycle Lane

Designated paths or lanes for cyclists at the sides of city streets are extremely obvious. They have to be, to encourage cyclists to use them and to make motorists aware of them, and they are demarcated in a range of different ways. In some instances the cycle lane is physically separated from the main carriageway by bollards or railings, but in most situations solid white lines at the sides indicate that motorized traffic is excluded while broken white lines show that vehicles can cross the cycle lane. Cycle-only signs are posted beside the lane, cycle symbols are often painted on the lane, and the surface of the cycle lane itself may be colored, usually red or green, to reinforce the message. These colored surfaces are usually designed to provide extra grip and reduce the chance of bicycles skidding on the wet pavement.

In some locations, colored surfacing is used only on those sections of the cycle lane where motorized traffic is most likely to cross, usually road junctions and entranceways, for these are the accident hotspots.

The Downside

While most cyclists agree that well-designed designated lanes—well marked, sufficiently wide, and thoughtfully routed—do increase safety for non-motorized road users, there are those who claim that restricting cyclists to the side of the road actually increases the risk of accidents at junctions. Indeed, many cities have traffic lights dedicated to cycle traffic for this reason. The use of cycle lanes on large roundabouts has largely been dropped because of the danger to cyclists.

Motorists, on the other hand, bemoan the loss of parking spaces that occurs as a result of implementing cycle lanes, while storekeepers and restaurateurs claim that reduced parking means fewer customers.

Off-Road Bike Path

With all the environmental and health benefits that cycling offers, it's hardly surprising that many municipalities have cycling policies and actively encourage the use of bicycles by creating cycleways separate trails that are closed to motorized traffic. In some cases these are part of statewide, and even nationwide, networks that partly utilize existing roads. Such systems, within towns and cities and in rural areas, tend to be routed through green spaces and along disused railroads, canal towpaths etc., but some run alongside major highways. These cycle paths, which are generally marked by similar signage to that used for on-road cycle lanes, often have a very smooth, high-quality surface. They are frequently shared by pedestrians or include a separate pedestrian path.

BICYCLE PARADISE

The Netherlands has a deserved reputation for being flat and having a great many bicycles, so it is no surprise that the country boasts an entire network of cycle paths that is largely independent of the road system. This network, which includes more than 20 long-distance routes totaling some 3,700 miles (6,000 km), even has its own traffic lights and directional signage. If there is a downside for the cyclist, it is that many main roads, bridges, and tunnels are closed to bicycles, but there's always another way round, and probably one that's a great deal more picturesque.

Stenciled markings and a painted surface denote a bicycle lane on this cobbled street.

MODERN ROUNDABOUT

WHAT?

A one-way rotary traffic system located at a road junction to facilitate the movement of traffic coming from several directions.

WHERE?

A common feature in Europe and Australia, roundabouts are increasingly being used in the USA, Canada, New Zealand, South Africa, China, and Israel.

DIMENSIONS

Ranging from mini-roundabouts with an outside diameter of 50 ft (15 m) to multilane, multilevel roundabouts more than 500 ft (150 m) across.

At junctions where intersecting roads need to be given more or less equal priority, roundabouts have been proven to have many advantages over uncontrolled junctions, traffic lights, or the North American four-way junction. They can handle more traffic, cause fewer delays, and reduce the number of automobile accidents. In some countries they are a relatively new feature on the transport landscape, but they are growing in number and spreading around the globe.

Roundabouts of the form that we see today first made their appearance in Britain in the 1920s. At about the same time, traffic circles were built in France and in the USA, but these differed from the modern roundabout in several fundamental ways. They were much larger, the traffic generally moved more quickly, and, importantly, drivers entering the circle had priority over those who were traveling round it. The opposite was true in Britain, where approaching traffic had to yield to traffic already on the roundabout.

Gridlock

The downside of the French and American system was that during periods of peak traffic—and traffic in the 20s and 30s was increasing exponentially—the circle would simply lock up. Traffic on one section of the circle was blocked by traffic coming in from the right, which was stationary because it was blocked by traffic coming in from the right, which was stationary…you get the picture. The outcome was a general disenchantment with traffic circles in the USA and, rather than changing the rules of priority, highways departments largely ceased to build them. The French persevered and gradually introduced yield signs at the entries to some traffic circles while others remained as they were, causing some confusion and a good many accidents, especially for visiting English drivers who already had enough trouble traveling the "wrong" way round the circle. In the last 20 years, France has moved almost entirely to the British style of roundabout and now has about half of the world's total number.

The Modern Solution

Since the 1960s, when many of the circular intersections in Britain were re-engineered to improve safety and handle greater traffic volume, and the mini-roundabout came into existence, roundabouts have proliferated throughout Europe and Australia. Many other countries have also begun to appreciate the roundabout's qualities, including cost (less expensive than traffic lights), safety (reduced risk of accidents for motorists and pedestrians), and smooth traffic flow, and over the last 20 years North America, too, has put aside its earlier bad experience and adopted the modern roundabout.

Design and Practice

The typical roundabout has a raised central island surrounded by a low-profile "mountable" apron—a ring around the island over which the wheels of large semi trailers can ride without causing or receiving damage. Roundabouts also have "deflection," a curve in the approach that forces traffic to slow down as it reaches the junction, increasing the chances of seeing oncoming traffic and generally reducing the speed at which vehicles circulate—a vital factor in the efficacy of roundabouts. Drivers are required to yield to traffic coming from the offside, to wait for a gap in the flow, and—on a multilane roundabout—to choose a lane and stay in it until they turn off, rather than weaving between lanes, which is what happened with the earlier design of traffic circle.

The only road users who are not entirely happy with roundabouts are cyclists, for whom the accident rates seem to be higher than at other forms of intersection, mainly due to drivers entering and leaving the roundabout failing to see them.

Organized Chaos

The winner of several unenviable accolades, such as one of the world's worst junctions and one of the UK's top ten scariest, the Magic Roundabout in Swindon consists of a circle of clockwise mini-roundabouts surrounding a central island around which traffic moves counterclockwise, all surrounded by an outer clockwise loop. The layout, designed by Frank Blackmore (the inventor of the mini-roundabout), allows a driver entering from the bottom of the diagram below and wanting to exit at the top right to choose between turning left and going round the outside or going straight ahead and passing to the right of the central island, which seems counter-intuitive. Nonetheless, the system has been in place for 30 years and it seems to work.

MAGIC ROUNDABOUT

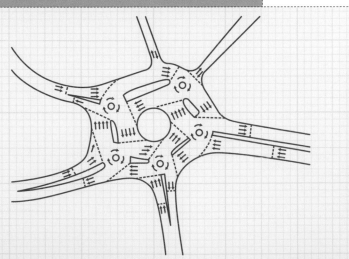

Not the kind of feature you would want to encounter on your driving test, Swindon's Magic Roundabout consists of five separate mini-roundabouts encircling a central hub. In a country that drives on the left, motorists can have difficulty with the idea of going counterclockwise around the center.

TRAFFIC SIGNALS

WHAT?
A signaling device usually comprising three separate lights (red, yellow, green), housed within an aluminum or polycarbonate casing and hung horizontally or vertically.

WHERE?
Normally found in groups at road intersections and displayed either at the side or above the roadway.

DIMENSIONS
There are two standard lens sizes: 8 in (20 cm) and 12 in (30 cm). The casing usually measures around 6 ft (1.75 m).

While the particulars of traffic signal, or light, displays vary from country to country, the standard three-color system has become a global icon of traffic management, and drivers the world over all stop on red and go on green. Having evolved from established railroad and shipping protocols, the pattern is so prevalent it is now applied to nonvehicular areas such as nutritional value and energy effectiveness.

The world's first traffic signal—a hand-operated set of semaphore arms controlling horse and carriage flow—was installed by J.P. Knight outside the Houses of Parliament, London, in 1868. Subsequent inventions and patents followed, including a flurry of more than 60 in the USA alone during the 1910s and '20s, from which emerged the electrically operated model familiar today.

Traditionally, three bulbs of up to 150W provided the light, but since the late 1990s these have increasingly been replaced with geometrically arranged arrays of LEDs (light emitting diodes). LEDs cost more but last longer, contrast better against direct sunlight, and switch faster, and the loss of one has little impact on the efficiency of the light as a whole. Moreover, replacing bulbs with LEDs is relatively easy—no new housing is required, meaning work can be carried out with minimal disruption. There is, however, one

drawback in cold regions: LEDs don't melt snow that may have accumulated as effectively as the big bulbs.

Light Control
Depending on its size and requirements, an intersection may need up to 20 separate traffic-light faces. The addition of pedestrian crossings, lane filters, and vehicle sensors further complicates matters.

A nearby metal box (usually the size of a small refrigerator) houses the controller that operates the signals and, by monitoring the electrical currents flowing through the lamp wires, checks for any conflicts (e.g. two green lights given for two potentially colliding lanes of traffic). Many cities have an additional central command post that can alter the timing of signals according to local needs. Various technologies can detect vehicles at intersections, enabling a smoother traffic flow. The most common is an induction loop, a coil of wire buried into the road surface that picks up the metal of a waiting vehicle above. This loop sends a signal to the controller cabinet, and the lights can change accordingly. More sophisticated models detect bicycles and motorbikes, while some sensors mounted on overhead poles detect strobes from oncoming emergency vehicles and make an immediate switch to green.

Intersections without lights are usually observed over a period of time to assess suitability. Criteria such as traffic and pedestrian volume, the number of crashes over a 12-month period, the proximity of schools, the effectiveness of signage, and the impact on the wider roadway network flow are all taken into account.

THE LAST STOP

"Shared space" is a concept fast gaining currency around the world. Established by Dutch traffic engineer Hans Mondeman, it favors human-based solutions to urban traffic problems. The use of traffic lights, together with road markings and signage, it argues, creates a form of control that robs each road-user of any sense of responsibility. By removing such paraphernalia, and by building environments that are sympathetic to the demands of all, people instinctively look out for themselves and each other. Recent schemes, whereby roads are shared and junctions go unmarked and unregulated, have not resulted in the carnage one may expect. Spot city traffic lights while you can—they may have reached the end of the road.

THREE-LIGHT TRAFFIC SIGNAL

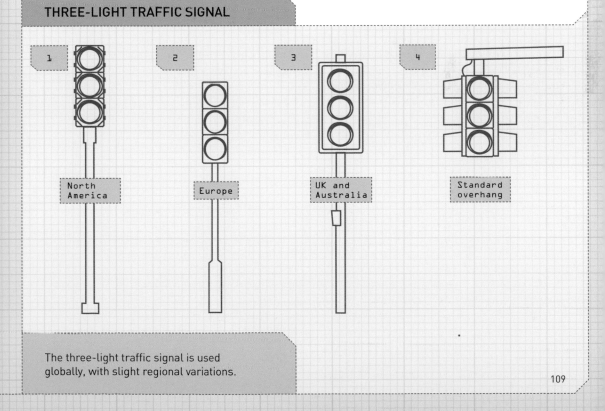

1 North America

2 Europe

3 UK and Australia

4 Standard overhang

The three-light traffic signal is used globally, with slight regional variations.

HIGHWAY

WHAT?

A major and well-constructed road having more than one lane in each direction and capable of carrying high-speed, high-density traffic.

WHERE?

Highways are found in virtually every country, but by far the most dense network is that of the USA.

DIMENSIONS

The Trans-Canada Highway runs from the Atlantic to the Pacific, a distance of almost 5,000 miles (8,000 km), and Australia's Highway 1, which runs all the way round the continent, is some 9,000 miles (14,500 km) long.

During the first half of the 20th century, the number of automobiles and the speed at which they could travel rapidly overwhelmed the capacity of existing roads around the world. More roads were needed, but the new roads had to be bigger, faster, safer, and integrated into a network. Programs to build systems of highways, and ultimately superhighways, were put in place, and these are the high-speed roads we drive on today to travel between, and even through, major towns and cities.

From the early 1900s on, a range of improvements were being made. As we have seen, there was progress in the methods of road construction and surfacing (see p.100), but the roads themselves were being built wider and with more gentle curves to facilitate cars traveling at higher speeds. Soon there were roads with two lanes going in each direction, with faster cars using the lane nearer to the center of the road.

Dual Carriageway

With relatively fast traffic moving in opposite directions, it made sense to place a physical barrier between them. One of the first to have dual carriageway on some sections was the Vanderbilt Parkway, which opened in 1908 and ran along Long Island, New York, from Suffolk County to Queens. The road, conceived by William K. Vanderbilt Jr.

as a race track, had banked curves and a concrete surface, and used bridges and overpasses to avoid "at-grade crossings," intersections where roads meet at the same level (see p. 114). It was the first road with access restricted to motorized vehicles only—no horses, no bicycles. Ironically, remaining parts of it are used as bike trails.

The idea of the divided highway, with traffic traveling in opposite directions being separated by a grass strip or by concrete or steel railings, gradually caught on, and several such roads were built in the USA, in Canada, and in Europe during the 1920s and '30s, generally to relieve traffic congestion within cities.

The First Superhighways

Over the same period of time, another trend was developing—the idea of a nationwide network of broad, high-speed highways running through rural areas and linking major cities. The Italians were the first to build one of these long-distance super-roads, in the 1920s. The 80-mile (130-km) road from Milan to Varese had only one lane in each direction, but it was the beginning of the Italian "Autostrada" expressway system. The network, consisting of some 4,000 miles (6,400 km) of multilane dual carriageway, now covers the whole country.

The German Autobahn system also started life in the 1920s, and expanded rapidly in the '30s with the enthusiastic support of Adolf Hitler and the toil of some 100,000 construction workers. By the start of World War II in 1939, a network of some 2,000 miles (3,300 km) of high-quality highway was in place, an asset that proved extremely useful when it came to moving troops and military hardware, and even landing airplanes. The modern Autobahn system has a worldwide reputation, not only for its quality and its total length (more than 7,600 miles [12,000 km], making it the world's third longest network), but also because it has no general speed limit, allowing drivers to put their foot to the floor on some parts of the system if road and traffic conditions permit.

The US Interstate Highway System

The idea of an integrated national system of highways throughout the USA was first put forward in 1939, but the project did not get underway until the 1950s. President Dwight Eisenhower, who had seen and appreciated the German Autobahn system and saw the benefits of such a project for road safety and for national economic development, supported the idea and in 1956 the Federal-Aid Highway Act established the program for funding and building it. At that time, the cost of constructing the projected 40,000 miles (64,000 km) of highway was estimated at $27.2 billion, and it was expected to take 12 years to build.

DETAILED SPECIFICATION

Highways accommodate high volumes of traffic traveling at high speed, and most countries therefore implement strict rules governing their construction. All freeways in the US Interstate Highway system, for example, conform to a set of standards that includes the following:

- Access is controlled, with points of entry limited to interchanges with grade separation.
- Interchanges are spaced a least 1 mile (1.6 km) apart in urban areas and 3 miles (4.8 km) apart in rural areas.
- All overpasses have a 16-ft (4.8-m) vertical clearance above the freeway.
- Interstates have at least two 12-ft (3.7-m) lanes of traffic in each direction.
- Right shoulders are at least 10 ft (3 m) wide; left shoulders are at least 4 ft (1.2 m) wide.
- Median width is 36 ft (11 m) in rural areas and 10 ft (3 m) in mountainous or urban areas (unless there are guardrails on the median).
- Design speed is generally 70 mph (112 km/h) (except in hilly terrain and some urban areas).
- No at-grade railroad crossings are permitted.

ANATOMY OF A SUPERHIGHWAY

Motorway, expressway, freeway, autobahn—the names by which different countries know their superhighways differ, as do the exact specifications, but all of them have certain features in common. The prime purpose of a superhighway is to facilitate safe, rapid, uninterrupted travel for heavy traffic, and this can only be achieved with these features in place.

Crossroads
Where other routes cross a freeway they do so as an underpass or overpass. Although highways and expressways may have traffic lights, freeways in general do not.

Signage (see p. 118)
Freeway directional signs are clear and highly visible from a distance. The frequency of signage on superhighways is greater than on other types of road.

Limited access
For safety, access is limited to vehicles capable of a certain minimum speed. This generally excludes pedestrians, horses and horse-drawn vehicles, bicycles, and small motorcycles. Some road systems require drivers to have a full license.

Roadside telephone
Some superhighways, such as the British motorway system, have regularly spaced telephones for use in the event of an emergency.

Multiple lanes
Almost all freeways have at least two lanes in each direction. Some have many more.

Road surface
Provides maximum grip in wet and wintry conditions

Interchange (see p. 114)
Entrances and exits take the form of ramps linking grade-separated routes so that traffic joining or leaving the freeway does not slow the traffic flow.

Central divide
Traffic moving in opposite directions is separated by a wide median or by a physical barrier on the central reservation.

Officially completed in 1991, 23 years later than expected and at a cost of $114 billion, the Dwight D. Eisenhower National System of Interstate and Defense Highways, as it has been known since 1990, now covers every state in the USA and connects every major city.

Not So Free Way

One very useful aspect of a freeway system is that access to it and exit from it can be controlled, and this makes it possible to charge motorists for the convenience of using it. In many countries certain sections of the freeway system are toll roads, or turnpikes, but in the case of France almost the whole of the 7,500-mile (12,000-km) "Autoroute" network is pay-as-you-go. Although the roads themselves are owned by the state, they were mostly built by private companies that retain concessions and charge for their use.

WORLD TOP TEN COUNTRIES BY LARGEST ROAD NETWORK

Country	Approximate Total Road Length	
	Miles	Kilometers
USA	3,958,150	6,370,031
India	2,062,730	3,319,644
Brazil	1,230,310	1,980,000
China	869,920	1,400,000
Japan	715,950	1,152,207
Russia	591,550	952,000
Australia	567,300	913,000
Canada	560,400	901,902
France	554,800	892,900
Italy	415,500	668,669

Some argue that expanding road capacity has the effect of increasing the amount of traffic, a phenomenon known as induced demand.

HIGHWAY INTERCHANGE

WHAT?

Element of a road system that allows traffic to move freely between two or more high-speed highways.

WHERE?

On every superhighway.

DIMENSIONS

The larger the interchange, the freer the flow of traffic; some cover more than a square mile (2.6 km²) and comprise more than 50 lane-miles (80 lane-km).

"Access must be controlled, with points of entry limited to interchanges with grade separation." This requirement is one of the defining qualities of a superhighway. It means, firstly, that you can't just drive onto the highway where you wish, even if your property is right beside it. You have to go to an interchange. Secondly, grade separation means that no roads ever cross a freeway on the same level and that traffic entering or leaving a freeway can do so without affecting the flow of traffic on the main route—ideally.

Roundabouts and traffic lights are effective ways to control traffic where roads intersect, allowing vehicles safely to cross or turn at the junction, but they necessitate slowing down or stopping, and this defeats the purpose of a freeway. Interchanges solve the problem by positioning separate highways on different levels and connecting them with ramps so that traffic can leave and enter at approximately the same speed as the main flow. Designs vary widely, but they are generally variations on, or hybrid combinations of, a few basic types.

RAMP TERMINOLOGY

When you come to a T-junction and you want to turn left, you turn left, but on a freeway interchange things are not so simple. In countries that drive on the right, it is rare for a left turn to be made from the fast lane—you have to turn off to the right initially, whichever way you want to go. There are three terms used to describe ramps on interchanges, and these express the relationship between the direction you want to take and the direction the ramp actually takes you.

- A directional ramp takes you the way you want to go.
- A non-directional ramp takes you the opposite way to the way you want to go—for example, it takes you round to the right in order for you to turn left.
- A semi-directional ramp starts by taking you the "wrong" way and then takes you the "right" way.

TYPES OF INTERCHANGES

Roundabout Interchange

A design that is commonly used in Britain where a minor road passes above or below a motorway is a roundabout placed at the level of the minor road. Traffic on the minor road has to slow down, but traffic on the motorway slip roads can move quickly.

Three-stacked Roundabout

Where two motorways cross, the same idea has been used but with both roads grade-separated from the roundabout. The design is compact, which is good where land is expensive, but the volume of traffic using such a junction can soon outstrip its capacity, forcing planners to introduce traffic lights on the roundabout and slowing the junction down.

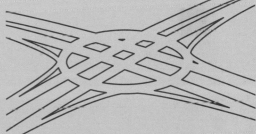

Diamond Interchange

This is the most common, and the simplest, form of freeway/minor road interchange in the USA. Its limitation is that where the ramps from the freeway meet the minor road there is a "normal" junction, requiring a Stop sign or traffic lights, which reduce traffic flow.

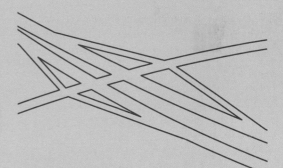

Dumbbell Interchange

This problem is reduced by replacing the ramp/minor road junctions with a roundabout on each side of the freeway, creating what is known as a dumbbell interchange because of its shape.

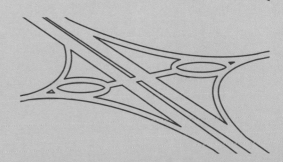

TYPES OF INTERCHANGES [CONT'D]

Trumpet Interchange

Used at three-way freeway intersections, the trumpet has been in use since the 1920s. It takes up a good deal of space but has fairly low construction costs and enables a free flow of traffic. It has some elements of the cloverleaf interchange (see opposite) and, like the cloverleaf, its loop forms a slightly confusing non-directional ramp.

Directional T Interchange

A common design at three-way junctions in towns in the USA, especially in New England, the Directional T is one of the few interchanges in which all the ramps are directional. The downside is that this can only be achieved by having left-turning traffic departing from the fast lane and merging traffic entering from the left (in a drive-on-the-right situation).

Semi-Directional T

Here the exiting and entering problem is solved by having two of the ramps crossing over or under the "bar" of the T, so that the turnoff lane becomes semi-directional. Nonetheless, this traffic still merges with the "upright" of the T from the left.

Triangle

In this variation of the T Interchange, the left turnoff ramp from the through route passes over the right turn ramp from the other carriageway and merges from the right.

TYPES OF INTERCHANGES [CONT'D]

Cloverleaf Interchange

In the USA and parts of Europe, this four-way interchange has long been the main way of dealing with intersecting freeways, but its popularity is declining for several reasons. The loops take up a great deal of space and can be disorienting. More importantly, "weaving" becomes a problem as traffic coming onto the freeway from a loop tries to move into the main lanes.

Four-Level Stack

Covering a large area, extremely expensive to build, and visually obtrusive, with traffic traveling on four levels, this is nonetheless the most effective interchange in terms of traffic speed and volume. The High Five Interchange in Dallas, Texas, even has a fifth level specifically for HOVs (high occupancy vehicles).

Whirlpool Interchange

Less expensive than the four-level stack and almost as effective, the whirlpool, or turbine, interchange has fewer levels but requires more space.

TRAFFIC SIGNS

WHAT?
Information and instructions for road users conveying messages in words and/or symbols.

WHERE?
Throughout every road network.

DIMENSIONS
Individual warning or information signs may be no bigger than a TV screen, but direction signs above major interchanges can be as wide as several lanes.

If you're driving on a paved road, whether it's in the Sahara or the Arctic, you are guaranteed to see signs beside or above it telling you what to do, what not to do, where to go, and what to watch out for. On any but the most minor route, even the surface of the road itself is covered in decipherable markings. Traffic signs vary considerably around the world, and yet most of them are remarkably easy to understand.

Almost every aspect of every road sign—its size, shape, color, position, height, symbol, typeface, even the angle at which it faces the road—is meticulously defined in an official document somewhere. Countries create their own symbols and wording, but most follow one of two broad systems—the North American or the European—that determine the general shape and color of particular categories of sign. The North American system is used in the USA and Canada, but also in Australia, New Zealand, and parts of Central and South America. The European system is based on the Vienna Convention on Road Signs and Signals, to which more than 50 countries have signed up. These include most of Europe, the Russian Federation, India, Pakistan, and parts of the Far East, Middle East, and Africa.

There are many categories of sign, but here are some of the main ones that you will see as you travel.

Regulatory Signs

These tell you what you must and must not do, and road users are legally obliged to obey them. Two important signs stand out from the crowd; these are the Stop sign and the Yield, or Give Way, sign. Both are vital to prevent an immediate accident, so it's a good thing they are the same in both systems—a red octagon for the Stop sign and an inverted triangle for the Yield sign.

In North America, other regulatory signs, such as speed limit signs, are generally vertical rectangles or squares with black messages on a white background or the reverse, although some,

WATCH OUT FOR WILDLIFE

In many parts of the world, highways cut through the migratory routes and feeding areas of many species of animal. Although wildlife warning signs are a serious matter (hitting a half-ton moose at speed can be lethal for both the moose and the driver), we seem to have a soft spot for silhouette depictions of animals, especially the more unexpected and exotic. These signs find their way onto everything from greetings cards to fridge magnets, but perhaps we should pay more attention to them beside the road.

such as No Right Turn, bear a circular decal. In the European system regulatory signs are red circles with black writing.

Warning Signs

Motorists need to know about any hazards that may be up ahead—railroad crossings or animals that are likely to be on the carriageway, for example—or

changes in the road, such as roundabouts and junctions. The American system generally uses yellow diamond-shaped signs with black written messages or symbols. Temporary warning signs, for example for construction or maintenance work in progress,

are generally the same as standard warning signs except they have an orange background. The Vienna Convention calls for a red triangle with black writing or a black symbol for warning signs.

Guide Signs

Many, if not most, of the signs along our roadways tell the motorist which direction to take to reach a particular destination, as well as how far away it is.

In the USA, these signs are generally a horizontal green rectangle bearing a white message. In the European system, there are distinctions between various types of road. Non-primary roads often have white rectangular directional signs with black writing, while primary roads have signs similar to those in the North American system—white writing on a green background with road numbers shown in yellow. The Vienna Convention states that motorways, autobahns, etc., can have signage either in white on green or white on blue. Britain is one of the countries that has opted for blue on all motorway signage and for references to motorway routes on other road signs.

CANAL

WHAT?
Navigable manmade inland waterway used for transport.

WHERE?
Mainly in North America and Europe, but also in China, Egypt, and Panama.

DIMENSIONS
Many canals are just a few miles long and no wider than two narrow barges, but the world's longest, the Grand Canal of China, runs for almost 1,120 miles (1,800 km), and the widest, the Cape Cod Canal in Massachusetts, is 480 ft (146 m) across.

The building of canals in order to ship raw materials and goods cheaply and efficiently helped to fuel the Industrial Revolution in both Britain and the USA, and though the great age of canals is long gone, artificial waterways remain a vital link in the world's transport system. Restored and maintained, many of the smaller canals also provide recreation and a fascinating connection with a past that has shaped our industrial landscape.

Both the Chinese and the Romans were great canal builders more than 2,000 years ago, but it was not until the 18th century that extensive and integrated canal systems began to be built. Many canals from the heyday of inland water transport can still be seen and admired today.

The Great Connectors
The Industrial Revolution, which began in England in the mid-18th century, created a demand for large quantities of raw materials, especially coal to fuel the new steam-powered factories, and it also generated the wealth to create new transport infrastructure. Roads at the time were rough and slow for horse-drawn carts, but on a canal a single horse towing a barge could pull the equivalent of 50 wagonloads. Beginning with links between coal mines and the growing industrial cities, a boom in canal building led

to the creation of more than 100 canals in Britain by 1820. The advent of the railways (see p. 128) soon led to their decline, but much of the canal system can still be seen throughout the English Midlands, restored and widely used by pleasure craft.

Canals in North America
Throughout the northeastern United States, and especially New England, there is still a network of small canals that in many cases were built to connect rivers to the water-powered factories in the 19th century, as well as being used for transport. As in England, these were almost all built by and for private owners.

The first major public canal project in the USA was the Eerie Canal, running from Albany on the Hudson River to Buffalo on Lake Erie, a distance of some 360 miles (580 km). Completed in 1825, this major work of civil engineering connected the Atlantic to the Great Lakes and it was an immediate success, opening up the states of the Midwest, carrying merchandise to and fro, bringing iron ore from Minnesota to the east coast, and boosting the importance of New York as a port. As part of what is now the New York State Canal System, more than 500 miles (800 km) of canals linking lakes and waterways across the state, the canal is still in use, mainly by pleasure craft but also by an increasing volume of commercial shipping.

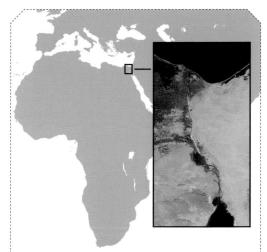

Canals are increasingly popular with recreational boaters.

In Canada, too, a major canal project has helped to shape the nation. The St Lawrence Seaway includes a sequence of channels and locks (see p. 122) that enable shipping to bypass unnavigable parts of the St Lawrence River and sail from the Atlantic all the way to Lake Superior. From there, the Welland Canal connects to Lake Erie, bypassing Niagara Falls.

OCEAN TO OCEAN

Canal transport is alive and well in many countries, but the world's most vibrant canals are those that connect seas or oceans. The Kiel Canal in Germany, completed in 1895, connects the Baltic and the North Sea, knocking some 200 miles (320 km) off the journey. The effect of the 120-mile (190-km) Suez Canal, which opened in 1869 and connects the Mediterranean and the Red Sea, is even more dramatic, as it cuts out a journey all the way round Africa.

NEW YORK STATE CANAL SYSTEM

The building of the Erie Canal advanced the settlement of the region, facilitated the exploitation of the region's resources, and helped New York City achieve its status as a major port.

CANAL LOCK

WHAT?

A mechanism to allow boats to move between levels in a canal system.

WHERE?

On most of the world's canals, at points where the canal must go through a change in level.

DIMENSIONS

Long enough to accommodate the ships using the canal; the world's largest, in Antwerp, Belgium, is 1,640 ft (500 m) long.

Canals that traverse level terrain take the form of a direct channel between two points, as the Suez Canal does, but most canals must deal with changes in level, either because they cross hills or because they link bodies of water at different heights. In almost all cases, the problem is solved by using a lock or a series of locks to raise or lower boats as they pass through the system. Locks can be seen on almost any canal, and they form a picturesque element on older and narrower canals.

The great Chinese canals initially used a mechanism known as a flash lock. A dam was built across the canal (or, in some cases, a river) between level sections, and when a boat needed to move downstream, a gate in the dam was opened to allow a rush of water to carry the boat through. Boats heading upstream had to be towed or winched through with the gates open, but the system was both dangerous and extremely wasteful of water, especially if the head of water was also needed to power a water mill.

The Pound Lock

The system that evolved from this, and the one that can still be seen in operation on navigable waterways worldwide, is called the pound lock. This consists of paired sets of watertight gates that enclose a basin, or pound, in which the water level can be allowed to rise and fall by letting water in from above or out to the level below through a valve at the level of the canal bed, usually a flat panel called a paddle or wicket that can be raised and lowered. On smaller locks this is done by hand using a rack and pinion that is turned using a windlass positioned beside the lock gates.

A lock basin allows boats to moor and to pass each other.

HOW A CANAL LOCK WORKS

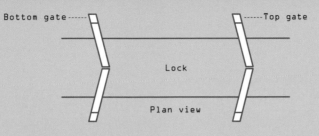

Plan view

1. With the pound filled to the level of the upper section, the upper gate can be opened and a boat can enter from the top level.

2. The upper gate is then closed, the paddle at the bottom gate is opened, and water is allowed to drain out into the lower part of the canal, causing the water level in the pound to fall.

3. When the water level in the pound is the same as that in the lower level of the canal, the lower paddle is closed, the lower gate is opened, and the boat continues on its way. A boat heading "upstream" can now enter the pound.

4. Once the lower gates have been closed, the paddle at the top gates can be opened and the pound can be filled from above, causing the water level to rise and the boat to rise with it.

5. Once the water is up to the level of the upper section, the top paddle is closed, the top gates are opened, and the boat moves into the upper section of the canal.

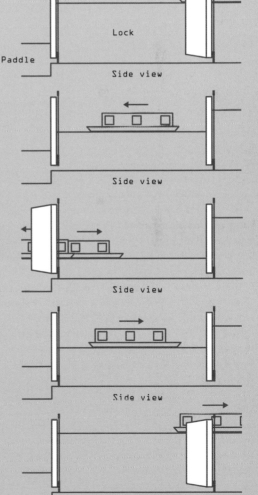

Miter Gates

The gates on early pound locks were flat and needed to be extremely strong and heavy to resist the enormous pressure of water bearing on them, making them difficult to open and close, but in the 15th century miter gates were invented in Italy. These work on a similar principle to that of an arch. The gates work in pairs, and rather than meeting to form a flat surface the two half-gates form an angle pointing upstream so that the pressure of the water holds them firmly closed against each other. With this design, the upper gates can only be opened when the pound is full, and the bottom ones can only be opened when it is empty, making it impossible for both sets of gates to be open at the same time (an event that would result in a huge loss of water). Small lock gates can be opened manually by pushing on a long counterbalancing arm, while the gates on large ship canals are operated by electric or hydraulic power. Some of the world's largest locks have sliding, rather than miter, gates.

Going Up Stairs

There is a practical limit to the rise that can be achieved with a single lock. When a greater change in height is required, a series, or flight, of locks is used, and these can be quite spectacular works of engineering. When a set of locks is continuous, the lower gate of one lock being the upper gate of the next, it is known as a staircase. A staircase of locks can be difficult to operate for the pilot of a boat moving "uphill," as each lock has to be full in order to fill the one below it before the ascent can begin. Neptune's Staircase, on the Caledonian Canal in Scotland, is Britain's longest staircase lock, a continuous flight of eight locks that lifts boats almost 65 ft (20 m).

When the topography allows it, locks are spread out along the canal, as they are at Caen Hill in southern England where, over a 2-mile (3.2-km) stretch, a series of 29 locks lifts the Kennet and Avon Canal almost 240 ft (72 m).

Boats can take up to 6 hours to ascend the Caen Hill Locks.

Conserving Water

While the pound lock is far more efficient than the flash lock in terms of wasting water, it does nonetheless use up a resource that can be in short supply, especially on the highest sections of a canal, which may receive very little water. On a small canal, a flight of locks can be replenished by pumping water from the bottom back to the top, but on large ship canals this is impossible. On the 48-mile (77-km) Panama Canal, which was opened in 1914 and connects the Pacific and the Atlantic, it is estimated that a ship uses 52 million gallons (196 million liters) of water to make its way through all the locks in the system. Now steps are being taken to conserve the resources of the Panama Canal watershed, including the building of new locks with a water-saving basin system that reutilizes much of the water used in each transit.

The Gatun Locks on the Panama Canal have double gates for security.

BOAT LIFTS

As an alternative to locks, some canals use mechanisms that lift the boats from one level to the next, either vertically or up a slope. The Morris Canal in New Jersey used water power to lift boats up a series of 23 inclined planes using trucks on rails, and both the South Hadley Canal in Massachusetts and the Chesapeake and Ohio Canal used similar systems.

The world's largest vertical lift, capable of raising ships of up to 1,488 tons (1,350 tonnes) in a tank, or caisson, can be seen on the Canale du Centre at Le Rœulx in Belgium, but the most spectacular one has to be the Falkirk Wheel in Scotland. This rotary boat lift has two counterbalancing caissons that are carried round a central axle, lifting boats 80 ft (24 m) from the Forth and Clyde Canal up to the Union Canal and enabling boats to travel from Edinburgh on one side of Scotland to Glasgow on the other. Toothed gears keep the two caissons horizontal as the mechanism turns.

FALKIRK WHEEL

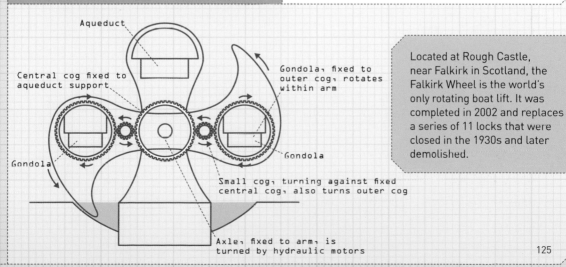

Aqueduct

Central cog fixed to aqueduct support

Gondola, fixed to outer cog, rotates within arm

Gondola

Gondola

Small cog, turning against fixed central cog, also turns outer cog

Axle, fixed to arm, is turned by hydraulic motors

Located at Rough Castle, near Falkirk in Scotland, the Falkirk Wheel is the world's only rotating boat lift. It was completed in 2002 and replaces a series of 11 locks that were closed in the 1930s and later demolished.

CANAL AQUEDUCT

WHAT?
Essentially a bridge carrying a navigable waterway.

WHERE?
In locations where a canal route needs to cross a relatively narrow valley, a river, road, or railroad.

DIMENSIONS
Most are less than 330 ft (100 m) long, but the world's longest, the Magdeburg Water Bridge across the River Elbe in Germany, is more than 3,000 ft (almost 920 m) in length.

Locks make it possible for a canal to follow a gradient, but when it comes to a steep-sided narrow valley, the practical solution is to cross it horizontally rather than going down and up again. Navigable aqueducts, or water bridges, have been around since the time of the Ancient Egyptians, but the great age of the canal brought new and aesthetic solutions that can be seen today, and raised waterways form an important part of the modern water transport system.

The great Canal du Midi, part of the French inland waterway system that connects the Mediterranean and the Atlantic, was completed in the 17th century and includes the first three modern aqueducts. All of them were built to cross streams and rivers, and the idea was later picked up in Britain when canal building began there in earnest. James Brindley, one of the engineers behind the Bridgewater Canal, designed the Barton Aqueduct to carry the canal across the valley of the River Irwell and when it was completed in 1761 it was hailed as an engineering marvel. Built of brick and masonry and lined with clay to carry the water channel, it was a large and heavy structure, but engineers such as Thomas Telford soon turned to cast iron as a lighter alternative. His Pontcysyllte Aqueduct (pictured above), a 1,000-ft (300-m)-long cast iron trough and towpath raised on stone piers that carries the Llangollen Canal 126 ft (38 m) above the River Dee in Wales, was built more than 200 years ago and it remains the highest aqueduct in Britain. The towpath takes the form of a shelf mounted along one side of the canal, allowing water to flow past the barge as it moves along the channel and reducing resistance. In 2009 it was made a UNESCO World Heritage Site.

Ever Longer
About 100 years later, in the 1890s, steel was used to build the impressive Briare aqueduct, which crosses the valley of the River Loire near Châtillon-sur-Loire in central France. Spanning 2,172 ft (662 m) and holding more than 14,300 tons (13,000 tonnes) of water, it was the longest navigable aqueduct in the world until 2003, when the Magdeburg Water Bridge in Germany was opened. With a total length of 3,012 ft (918 m), it crosses the Elbe River to connect two previously separate canals.

The covered aqueduct in Metamora, Indiana.

BARTON SWING AQUEDUCT

Covered Aqueduct

Metamora, in Indiana, calls itself Canal Town and boasts the United States' only covered aqueduct, carrying a remaining section of the Whitewater Canal over Duck Creek. The canal, which linked the agricultural Whitewater Valley and the Ohio River, was begun in 1836 and completed in 1847. Stretching 76 miles (120 km) from Hagerstown to Lawrenceburg, it descended almost 500 ft (150 m) and had 56 locks along its length.

When the channel of the River Irwell became part of the Manchester Ship Canal and larger sea-going ships were able to move through it, Brindley's Barton Aqueduct was too low to allow passage. It was replaced in 1893 by a remarkable one-of-a-kind combination of cantilever bridge, swing bridge, and aqueduct that now carries the Bridgewater Canal. The 330-ft (100-m) Barton Swing Aqueduct pivots about its center on an island in the middle of the canal and swings through 90 degrees to allow ships to pass on either side.

Built with local stone, the Dundas Aqueduct in southern England was completed in 1805.

RAILROAD TRACK

WHAT?

A network of steel rails over which locomotives pull and push passenger cars and freight cars.

WHERE?

Railroads play an important role in almost every country in the world.

DIMENSIONS

Approximately 850,000 miles (1,370,000 km) long, and commonly less than 5 ft (1.5 m) wide.

The first public railroad, 25 miles (40 km) of track on which Robert Stephenson's steam-powered "Locomotion No. 1" ran between Stockton and Darlington in northeast England, opened in 1825. By the time the railroad celebrated its 50th anniversary, there were some 160,000 miles (257,000 km) of railroad track worldwide. Today that figure is 850,000 miles (1,370,000 km). The world's railroads are not only an engineering marvel in their own right—they are also an ideal way to see the countries they traverse.

Just as the canals evolved to meet the needs of the Industrial Revolution, so did the railroads. In the coal mines, horses hauled wagons along rails, and these were then extended to the factories or to the nearest canal, but with the invention of the steam locomotive the railroad soon became the quickest and most cost-effective means of bulk transport. The importance of the canals diminished rapidly, and by the end of the 19th century networks of railroads connected the major towns and cities in North America, the countries of Western Europe, and British-controlled India. In the course of the 20th century, railroads have been constructed in almost every country in the world, carrying freight and passengers cheaply and efficiently.

Railroad Basics

Traditionally the railroad consists of a foundation of crushed stone, or ballast, on which railroad ties,

or sleepers, made of timber or prestressed concrete (see p. 26) are laid crosswise. The rails, which not only carry the steel-wheeled locomotives, passenger cars, and freight cars with the minimum amount of friction but also steer them, are fixed to these. Two types of rail are in use: the flat-bottom rail and the bullhead rail. Various kinds of clip are used to fix the rails to the ties, and there is usually a rubber cushion beneath the rail. The lengths of rail are joined to each other with plates or, in the case of high-speed rail, they are welded end to end.

Low Vibrations

In recent years, a new type of design known as slab track or non-ballasted track has come into use, especially in tunnels and locations where vibration needs to be kept to a minimum. In designs such as Sonneville's Low Vibration Track (LVT) system, the rails are attached to shock-

LONGEST AND FASTEST

Although it is hard to obtain completely up-to-date figures for many countries (and some are building railroads while others are closing track down), the top five longest rail networks are clear. Lengths have been rounded to the nearest thousand miles or kilometers.

1. **United States** 140,000 miles (226,000 km)
2. **Russia** 80,000 miles (128,000 km)
3. **People's Republic of China** 57,000 miles (91,000 km)
4. **India** 40,000 miles (64,000 km)
5. **Canada** 36,000 miles (57,000 km)

China recently announced the opening of the world's fastest regular passenger service. The bullet train, with a maximum speed of 262 mph (420 km/h), will cover the 126 miles (420 km) from Shanghai to Hangzhou at an average speed of 220 mph (354 km/h).

absorbing units in a level trough and concrete is then poured around these.

Staying on Track

It is critical that the rails be absolutely parallel—ideally to within 0.04 in (1 mm) for high-speed trains. The distance between the inside edges of the rails is known as the gauge, and in the early days of rail the width of the track was a subject of considerable debate, balancing cost and cornering against stability and capacity.

George Stephenson chose 4 ft 8.5 in (1.435 m), the approximate wheel span of carts and carriages for thousands of years, as the gauge for the Liverpool and Manchester Railway (the first mainline railroad), while his railroad-building rival, Isombard Kingdom Brunel, gave his locomotives a wheel span of 7 ft (2.134 m).

Stephenson eventually won out and his track width, now known as Standard Gauge, is used by 60 percent of the world's railroads. None is now as wide as Brunel's, but India and Russia have gauges that are 5 ft (1.524 m) or wider. New Zealand, Japan, parts of the Far East and Australia, and central and southern Africa use gauges of 3 ft 6 in (1.067 m) or 3 ft 3.37 in (exactly 1 m).

On high-speed railroads, sections of rail are welded together to provide a smooth ride.

RAILROAD SIGNALS

WHAT?

A system of signals to inform locomotive drivers whether or not the track ahead is clear and whether they may proceed.

WHERE?

Alongside railroad tracks worldwide. The greatest numbers are found in the vicinity of busy railroad junctions, and especially around major rail stations.

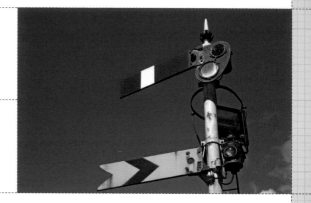

DIMENSIONS

Railroad signals range from low lights, almost at ground level, to tall posts, gantries, and signal bridges that extend above or across the railroad.

Railroad transport is fast and cost-efficient, but there is one thing that locomotives are very poor at—taking evasive action. They can't swerve and they can't stop quickly, especially when pulling a large number of wagons, so the only way to avoid collisions is to make sure two trains never share the same section of track. This—and warning of other hazards ahead—is what railroad signals the world over are designed to achieve.

SEMAPHORE SIGNALS

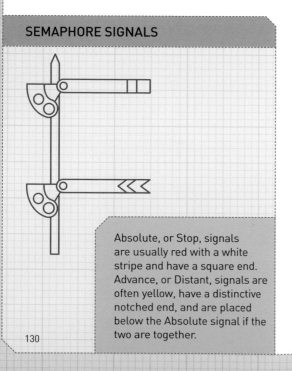

Absolute, or Stop, signals are usually red with a white stripe and have a square end. Advance, or Distant, signals are often yellow, have a distinctive notched end, and are placed below the Absolute signal if the two are together.

The fine detail of railroad signaling varies around the world, but the broad principles are generally the same. A railroad system is divided into lengths of rail track known as "blocks," and signals are used to prevent two trains from occupying the same block at the same time. The signals also tell the locomotive driver whether the subsequent block is occupied

The illuminated red on this two-aspect light signal indicates "Danger-Stop."

so that the speed of the train can be adjusted accordingly. Signals are also used to stop trains and warn of other forms of obstacle ahead.

Semaphore Signals

Formerly the most common form of railroad signal, the mechanical semaphore signal consists of a hinged arm that conveys information by its position and colored lenses attached to the arm that move into different positions in front of a light source. There are two forms of semaphore signal: lower quadrant and upper quadrant. A lower quadrant signal indicates Stop and shows a red light when it is horizontal, and indicates Clear, with a green light, when it is inclined downward at an angle of 45 degrees. The more common upper quadrant signal also indicates Stop and shows a red light when it is horizontal, but indicates Clear, with a green light, when it is inclined upward at an angle of 45 degrees. This is considered safer, because if the signal malfunctions and gravity causes it to fall to its resting position, it will signal Stop rather than Clear. "Absolute" or "Stop" signals—used to halt a train at the start of an occupied block, at a junction, or at an obstacle such as a swing bridge—generally have a red blade with a white stripe.

Advance, or Distant, signals, which warn that the next Absolute signal is in the Stop position, are generally yellow and have a green light for Clear and a yellow light for Caution. In some countries the Distant signal is red and is distinguished from the Absolute signal by having a chevron notch in its tip.

Light Signals

On most railroads, semaphore signals have been replaced by light signals. Colored light signals operate like a highway traffic signal, using red, green, and sometimes amber lights to indicate the status of the railroad ahead, while position light signals have several lights of the same color (usually white or amber) and use patterns of illuminated lights to convey information. Some railroads use both kinds separately, and some use a system that combines elements of both.

OPERATING THE SIGNALS

The earliest signals were operated by levers at the base of the signal post, and later by levers in an "interlocking tower" or "signal box" connected to the signal by cables or rods. The term "interlocking" refers to the fact that the signal and switch, or points, levers were interconnected in such a way that they could not contradict each other. All signals are now operated electrically, generally on low voltage so that batteries can be used to provide backup power. Manual operation has largely been replaced by automatic signaling systems in which electrical relays detect whether an electrical circuit in each of the rails within a "block" is being short-circuited by the metal wheels and axles of a train. The signals are adjusted accordingly.

Light signals have replaced mechanical signals in most countries.

RAILROAD CROSSING

WHAT?
An intersection between a roadway and a railroad where the two are at the same level.

WHERE?
Throughout every country that has a rail system. The USA has an estimated 140,000 at-grade railroad–public highway crossings and a further 87,000 on privately owned roads.

DIMENSIONS
At-grade crossings are not generally used where there are more than two sets of rail track or two lanes of road traffic in each direction.

One of the keys to the speed of freeway traffic is the absence of at-grade intersections. Railroads, on the other hand, with trains traveling at more than twice the speed and quite unable to stop in a short distance, do have roads crossing them at the same level. Little wonder, then, that railroad crossings are the focus of a host of safety features and yet remain the scene of many dramatic accidents and fatalities.

In the 19th century, when trains traveled slowly and the fastest road traffic was a horse, level crossings were already seen as a problem, and manually operated full-width gates were used to keep road users out when a train was due. When the gates were swung open for road traffic, they closed off the railroad to keep people and livestock off the track. As trains have become faster and more frequent, and road traffic has come to include freight trucks, school buses, gas tankers, and millions of automobiles, so the railroad crossing has had to evolve too.

Crossing Design
Railroad crossings themselves need special features to allow traffic to cross at all. The surface of the road needs to be as smooth as possible, and the rail tracks are therefore set into the ground instead of being on sleepers. A level crossing is a bad place to get stuck, so where there is a risk of grounding, signs may tell drivers of long low vehicles to phone the railroad authorities before crossing. On electrified railroads that use a third rail to carry the power, this rail has to stop well before the crossing and start again after it to avoid the risk of electrocution. When the current is carried by overhead cables, these need to be high

Lights and a barrier mark a railroad crossing in Italy.

enough to clear tall vehicles, and there are generally signs before the crossing to warn drivers about the safe height.

Barriers and Gates

At many crossings, full-width gates have given way to vertically hinged half-barriers. Some are operated remotely by railroad staff, but many fall and rise automatically, responding to sensors that detect the approaching and departing train. These railroad crossings also have flashing red lights and an audible warning that begin before the barriers come down and continue until the train has passed and the barriers are up. At some automatic level crossings, drivers of large or slow vehicles are required to phone before and after crossing.

Open Crossings

Many railroad crossings, and in some countries the vast majority of them, have no gates or barriers at all. Some are equipped with automatic traffic signals in the form of lights and bells to warn of oncoming trains, but others (almost half of those in the USA, and two-thirds of Australia's 9,000 crossings) simply indicate the presence of the railroad by means of signs and rely on drivers to take the necessary precautions.

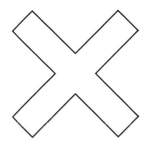

In the USA, the sign for an uncontrolled crossing is the crossbuck, a diagonal white cross bearing the words RAILROAD CROSSING. Variations on the crossbuck are found all around the globe, from the red-outlined white cross used in Canada, through the flattened and point-ended version found in Britain, to the red and yellow Finnish crossbuck.

IGNORING THE SIGNS

At-grade crossings are recognized by transport authorities as the greatest hazard on the railroad networks, and many hundreds of people are killed at railroad crossings every year. Despite all the technology, education, and enforcement, an incredible number of motorists are willing to risk their lives by nipping in front of an oncoming train—even zigzagging through half-barriers—to save a few minutes, or simply fail to see the warning signs. Generally the locomotives and railcars come through relatively unscathed, but level crossing collisions have derailed passenger trains with tragic results.

Variations on the crossbuck are found around the world.

RAIL FREIGHT YARD

WHAT?
A location at which freight-carrying railcars are loaded and unloaded, or sorted and redirected.

WHERE?
Rail freight yards are found at most major ports and close to many large cities, as well as at "hubs" where several railroads converge.

DIMENSIONS
A large freight yard can cover 1,000 acres (400 hectares) and comprise more than 100 miles (160 km) of rail track.

In terms of the energy costs, rail is by far the cheapest way to transport freight between two connected points. The downside of the system is its inflexibility—freight trains only run within the limits of the railroad system—and the consequent additional, and often heavy, costs for moving the freight between different modes of transport at the start and end of the journey, as well as between trains. It is the function of freight yards of various kinds to make this process of "transshipment" as efficient as possible.

As we have already seen (see p. 128), the railroads started life as a means of conveying coal from the mines to the seaports, and this kind of bulk transport—carrying coal, ore, aggregate, cement, etc., from point of origin to final destination—remains an important role for railroads worldwide. A railroad train hauling freight (often a single commodity) all destined for one location is known as a unit train or block train, and the larger the load and the greater the distance the more efficient it is. Car manufacturers, for example, use unit trains to carry cars from the factory to the shipping port, and Tropicana's refrigerated "Juice Trains" that run from Florida to New York State, Ohio, and California are cited as an example of the efficiencies of rail transport even for perishable goods.

Containers make up a large part of modern rail freight.

Container Freight Yard

For non-bulk commodities traveling relatively short distances, rail has conceded first place to road haulage, but the advent of container shipping (see p. 159) has kept rail in the running for long-distance transport. Most marine container terminals are also rail freight yards where containers are transshipped from sea-going vessels onto flatcars (or flat wagons, as they are known in Britain) using special gantry cranes. Some flatcars can carry two containers end to end, but there is a growing trend, especially in North America and Australia, toward double stacking—placing one container on top of another on railroads where there are no clearance issues caused by tunnels, bridges, or overhead electrical cables. Canadian National Railway runs double-stacked container freight trains almost 2.5 miles (4 km) long using distributed power—several locomotives positioned along the length of the train.

Classification Yard

With the exception of unit trains, the cars that make up a train are ultimately heading for different destinations and will have to be separated and redirected at some point. That point is a classification yard (or marshaling yard as it is known in Britain and parts of Canada).

A classification yard, which is generally located at a major intersection between two or more railroad networks, consists of many long parallel sections of rail track connected by switches, or points, to a "lead" track. The cars of a train are uncoupled and

THE WORLD'S LARGEST

Located between major east–west and north-south rail corridors in North Platte, Nebraska, Union Pacific's Bailey Classification Yard covers almost 4.5 square miles (11.5 km^2), comprises more than 300 miles (500 km^2) of rail track, and sorts some 3,000 railcars per day. Handling a significant proportion of the goods moving within the country, it has been described as a barometer of the US economy.

then sent individually along the lead, through the switches, and into particular tracks in order to make up trains of cars all heading for the same destination. The train can then be attached to a locomotive and sent on its way.

In a flat yard the cars are moved into position by shunting locomotives. Gravity yards are built on a slope so that the cars roll through the system without being pushed. The most common type of classification yard, though, is the hump yard, in which the lead passes over a small artificial hill. The cars are pushed to the hill and then released to roll down the other side. A "retarder" system automatically grips the sides of the steel wheels to ensure that heavy cars do not slam into those already positioned ahead.

Road and rail networks meet at the freight yard.

LIGHT RAIL SYSTEM

WHAT?
An urban passenger transport system that uses trams/trolley cars/streetcars running on rails and largely sharing the existing road system.

WHERE?
In a growing number of cities around the world.

DIMENSIONS
Street rail systems in many cities exceed a total track length of 60 miles (100 km).

Clean, quiet, and efficient light rail systems graced the streets of many cities in Europe and North America in the 19th and early 20th centuries but largely went out of favor in all but a few European countries as the automobile came to dominate the roads. In the last 30 years, however, street railways, or tramways as they are known in Britain, have made something of a comeback.

The first street railways were horse-powered and developed, as the railroads did, from rail haulage systems that benefited from the smooth ride and low friction afforded by steel rails and steel wheels. Horse cars, as they were known in the USA, became popular in many cities, but with the development of the steam locomotive the horse-drawn carriage gradually gave way to the steam tram. Toward the end of the 19th century, electricity became, and has remained, the power of choice for light rail systems.

POWERING LIGHT RAIL

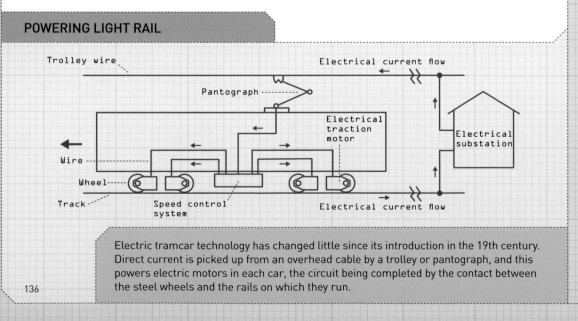

Electric tramcar technology has changed little since its introduction in the 19th century. Direct current is picked up from an overhead cable by a trolley or pantograph, and this powers electric motors in each car, the circuit being completed by the contact between the steel wheels and the rails on which they run.

THE CABLE CAR

With the mechanization of the streetcar came another alternative form of propulsion—the cable car, pulled by a system of cables and pulleys powered first by stationary steam engines and later by electricity. In general, the complexity of having moving cables running through the streets proved to be a major disadvantage, but the cable-pulled system—offering plenty of power and a steady speed—came into its own on steep hills, which helps to explain why San Francisco retains them to this day.

Going Electric

The availability of a reliable electrical supply led to a flurry of creative invention from the 1880s onward in the search to find ways of providing a moving vehicle with electricity. There were attempts at supplying power through the ground rails, but shocked horses and pedestrians were unimpressed, and a variety of overhead solutions were tried. The first successful method is attributed to Frank J. Sprague, the builder of the Richmond Union Passenger Railway in Virginia, USA, which opened in 1888. His system used a trolley, a grooved metal wheel at the end of a spring-loaded pole mounted on the roof of the "trolley car." The wheel ran against an electrical cable to provide the necessary contact, and the rails provided the "ground." In a later improvement, and one that is still used, a carbon shoe replaced the wheel, providing better contact and less wear on the electrical cable. The trolley pole has now been largely superseded by the pantograph, a sprung diamond or L-shaped structure that works better at higher speeds and makes the streetcar less likely to come "off its trolley."

Expansion and Contraction

Throughout the first half of the 20th century, tramcar systems made their mark in many cities around the world. The street railway network in

Buenos Aires, Argentina, for example, had more than 500 miles (800 km) of track until it was decommissioned in the 1960s. The Russian city of St Petersburg, which has had a street rail system since the 1880s, reached a peak of some 210 miles (340 km) of track, but has been dismantling its network over the last 15 years. At one time the tramway system in Sydney, Australia, rivaled that of London, carrying more than 400 million passengers in 1945, but the service was reduced in the 1950s and closed in 1961. Melbourne, on the other hand, has continued to keep street rail as an important, and extremely popular, form of public transport, and now has the largest network in the world, with more than 150 miles (240 km) of track.

In many Central and Eastern European countries the tram or streetcar has played a central role in public transport for more than a century, and Germany, too, has retained its street rail systems. In most of Western Europe, on the other hand, the networks fell from favor but since the 1980s there has been a resurgence in a modernized form of light rail, in which some parts of the rail system are segregated from the road system and have their own rights of way. In the USA, too, many cities are now bringing back the streetcars.

TUNNEL

WHAT?
Long underground excavations through which canals, railroads, rapid transit, and roadways can run.

WHERE?
Wherever going through is better than going round or over.

DIMENSIONS
When it is completed, the 35-mile (57-km) Gotthard Base Tunnel—twin tunnels to take the railroad under the Swiss Alps—will be the world's longest transport tunnel.

The first tunnels were built more than 2,000 years ago, but they were created as aqueducts to provide towns and cities with water rather than for transport (although the Ancient Roman road Via Flaminia includes a short tunnel excavated in the first century CE). It was the growth of the canal system and subsequently the railroads that made tunnel building a vital part of civil engineering.

Recognizing the costs, dangers, and difficulties of tunneling, the early canal builders chose to go around hills where they could and built lock systems to go over them where they could not. However, both solutions added time to the journey, and lock systems required large quantities of water, so inevitably there were situations where a tunnel was the best choice. The very first canal tunnel, the 540-ft (165-m) Malpas Tunnel, was excavated in 1679, in Herault, France, on the Canal du Midi.

From Water to Rail to Road
By the early 19th century, during the rapid expansion of the canal network in Britain, tunnels were becoming more common, and in 1811 the Standedge Tunnel on the Huddersfield Narrow Canal was opened. More than 3 miles (almost 5 km) long, it remains the longest canal tunnel in Britain. In the USA, the Union Canal Tunnel in Pennsylvania was built in the 1820s, and although the canal was closed

in 1881 the tunnel itself is now a National Historic Landmark. The Paw Paw Tunnel on the Chesapeake and Ohio Canal is over 3,000 ft (almost 1 km) long and was excavated through extremely difficult rock, taking 14 years to complete and opening in 1850. Throughout the 19th century, the growing railroad networks, especially in Europe and North America, also required tunnels, and tunneling techniques made great advances. Then, with the arrival of the automobile, it was the turn of the highways to go underground and underwater.

Cut and Cover
Useful in situations where the tunnel can be located close to the surface, such as some of the first underground rail tunnels to be built in London in the 19th century, this method involves excavating a channel in which the tunnel is constructed and then backfilling around the tunnel up to ground level. In cases where the surface needs to be reinstated quickly

(where, for example, a tunnel passes beneath a street), it is even possible to dig down on either side of the tunnel, pour concrete walls, and construct the roof of the tunnel, returning the surface to a usable state, and then excavate the tunnel beneath the roof. The floor is the last part of the tunnel to be constructed.

Bored Tunnels

The tunnels up until the middle of the 19th century were built using methods developed from mining, digging with hand tools or simple drilling equipment to excavate the rock and earth, which was removed using rail carts, shoring up the roof with timbers, and lining the tunnel with brickwork. Drilling and blasting methods were, and still are, used to excavate through hard rock, but this type of excavation has largely been superseded by tunnel-boring machines, or TBMs. These giant machines, up to 50 ft (15 m) in diameter and with a rotating head armed with disk cutters, can bore through any kind of rock with pinpoint accuracy.

UNDERWATER TUNNELS

Although many tunnels beneath water—for example, the Channel Tunnel between England and France—have been built in this way, boring is a difficult and dangerous method, especially through soft rock. Since 1940, a method known as "immersed tube" tunnel building has been used in a variety of underwater situations. This involves dredging a channel in the bed of the river or sea, sinking precast tunnel sections, and then joining them together underwater. A protective layer of gravel and rock is then laid over the tunnel. The first concrete tunnel of this kind was built in Rotterdam under the Maas estuary. The longest and deepest built so far is the 3.6-mile (5.8-km) steel Transbay Tube that carries BART (Bay Area Rapid Transit) across San Francisco Bay at a maximum depth of more than 130 ft (40 m). It was opened in 1974.

IMMERSED TUBE TUNNEL

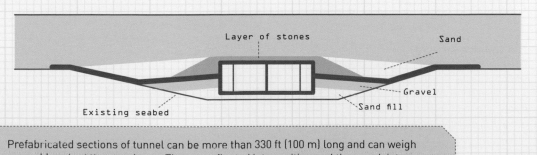

Layer of stones · Sand · Existing seabed · Sand fill · Gravel

Prefabricated sections of tunnel can be more than 330 ft (100 m) long and can weigh several hundred thousand tons. These are floated into position and then sunk into a pre-excavated trench. The sections are leveled, joined, and sealed together, the tunnel is buried in gravel, sand, and rock, and the seabed is reinstated.

RAPID TRANSIT SYSTEM

WHAT?
Also known as a subway, metro, or underground system, rapid transit is an electric urban passenger rail system operating largely in underground tunnels.

WHERE?
Some 160 cities around the world have rapid transit systems.

DIMENSIONS
The city with the world's longest subway system—264 miles (425 km) of passenger route—is Shanghai in China. The city with the most subway stations is New York with 423.

As well as inventing the trolley-pole used on streetcars and a constant speed electric motor, Frank J. Sprague also developed "multiple unit traction control," which allowed several electrically powered cars coupled together to be controlled in unison. Bring this together with advances in tunneling and a growing need for transport in cities whose streets are already crowded and you have—rapid transit.

Known in New York as the Subway, in London as the Underground or the Tube, in Paris, Tokyo, and Moscow as the Metro, and in Vancouver as the SkyTrain, rapid transit has much in common with light rail but has one major difference—it is "grade-separated" from other forms of transport. The subway train doesn't have to wait for pedestrians or road traffic to cross its path because it travels either beneath or above the city streets. This makes it faster and able to carry a higher volume of passengers.

The First Underground
Steam trains running in underground tunnels constructed by the cut-and-cover method were first used on London's Metropolitan Railways in the 1860s, but the first deep tunnel electric "tube train" was the City and South London Railway, which opened in 1890. When work began in 1886 on construction of the two 3.2-mile-long (5 km) tunnels

running under the River Thames, the idea was to pull the trains through using stationary engines and a cable system of a kind already in use in San Francisco (steam power being impractical in an under-river tunnel). However, the company supplying the system went broke two years later and the decision was taken to go electric, using a "live" third rail, making this the first major electric railway.

RT Goes Global
The railway had its financial problems, but as an experiment in technology it was clearly a success. Other underground railways in London soon changed over to the much cleaner electric locomotives, and the rapid transit concept was quickly taken up in several European cities including Budapest, Glasgow (where the system was cable-hauled until the 1930s), Paris, and Berlin (where some sections were elevated above street level).

In the USA, in 1897 the Green Line in Boston extended the existing light rail streetcar system to run through tunnels into the downtown (which it still does, carrying more passengers than any other light rail system in the USA). In 1904, the first section of the New York City Subway opened between City Hall and 145th Street, the first step in what was to become one of the world's largest rapid transit systems, now carrying some 5 million passengers every weekday.

Philadelphia and then Buenos Aires, Argentina, followed suit, and by the end of the 1930s subway systems were running in Tokyo and Osaka in Japan, and in Moscow.

Still Growing

Rapid transit has gone from strength to strength, providing a cost-effective solution to the needs of expanding cities where the density of traffic and the cost of real estate rule out surface options for new public transport. Since the 1950s, Metro systems have been built in cities throughout the then Soviet Union, in Toronto and Montreal in Canada, in several Brazilian cities, including Rio de Janeiro and Sao Paulo, and, since the 1980s, throughout China.

PRESERVING HISTORY

Underground rail systems inevitably reflect the style and architecture, even the mood, of the times in which they were built. The stations of the Moscow Metro, for example, built under the Communist regime as "palaces for the people," are famous for their ornate design, replete with marble columns, sculpted ceilings, and elegant chandeliers. The Montreal Metro, in which artists were invited to make major contributions to the stations in the form of murals, sculptures, and stained glass works of art, has been described as the world's most beautiful subway system, and Line A on the Buenos Aires Metro, or Subte, still retains its original wooden coaches, which are now more than 90 years old.

New York City's subway system dates back to the early 20th century.

BEAM BRIDGE

WHAT?

A bridge consisting of a simple platform with its ends resting on supporting abutments on either side of a channel.

WHERE?

Beam bridges are a common, and relatively inexpensive, solution for bridging short spans.

DIMENSIONS

Single spans seldom exceed 250 ft (75 m) but, with the use of intermediate supporting piers, a beam bridge can be any length.

The first bridge ever used was undoubtedly a beam bridge—probably a fallen tree across a narrow stream. Simple log or plank bridges are still used around the world to bridge short gaps, but the longer the span and the heavier the beam, the stronger that beam has to be. Over the centuries, engineers have come up with a host of ways to build in strength while keeping the weight to a minimum.

A beam bridge (also known as a "simply supported" bridge, because the weight of the bridge acts directly downward on its supports) needs to be stiff to resist the forces acting on it. The weight of the beam itself and any load placed on it tend to cause it to bend, compressing the top layer of the beam and stretching the bottom layer. An unevenly distributed load will also tend to cause the beam to twist, so it needs compressive, tensile, and torsional strength. The properties of wooden planks and stone slabs restricted their use to very short spans, but the development of materials such as iron and steel in the 18th century, and reinforced concrete in the 20th, opened the possibility of much longer spans.

Girder Construction

Overcoming the limitations of a single slab, girder construction involves the use of several structurally strong members running the length of the span, with the bridge surface, or "deck," laid on top of these. The girders were originally made of iron but later of steel and now commonly of prestressed concrete, combining the high tensile strength of steel with the compression qualities of concrete. The girders have a cross-section that gives them the greatest strength

A BEAM UNDER LOAD

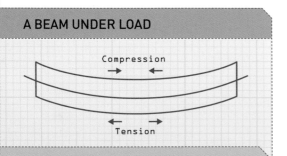

A load placed on a beam tends to cause it to bend, requiring the beam to withstand a compressive force in its upper layers and a tensional force in its lower layers.

for the least weight, such as an I or L shape when seen from the end. This design is light and relatively cheap, but while it works well for straight spans, it is less resistant to twisting and therefore doesn't lend itself so well to a curved deck of the kind seen in highway overpasses, a common use for beam bridges.

Box Girder

For even longer spans and for curved applications, box girder construction has still greater strength and is better able to withstand twisting forces. As the name implies, the deck is supported by one or more girders that have a boxlike form, usually rectangular or trapezoidal in cross-section. Prestressed and reinforced concrete box girders can be constructed on site using supporting "falsework," while steel box girders are usually prefabricated and then bolted or welded together on site.

This method of construction was particularly popular during the 1960s, but a series of bridge collapses in the early '70s led to a serious reconsideration of the method. The downside of the box girder is that the interior is difficult to maintain, and moisture seeping into reinforced concrete can cause the steel reinforcement to corrode, weakening the structure.

Continuous Span

Although the length of a single span is restricted by the properties of the materials, there is no limit to the number of spans that can be joined together using supporting piers. This method is used to create extended highway flyovers, such as the soaring Rodovia dos Imigrantes in Brazil, and multispan crossings over large extents of water, such as the Lake Pontchartrain Causeway in southern Louisiana, which is almost 24 miles (38.4 km) long.

NOT SO NEW

The concept of the box girder was pioneered in Britain back in the 1830s by engineer Sir William Fairbairn and mathematician Eaton Hodgkinson, using wrought iron plates. Fairbairn used this method in the design of a crane jib, in which the top surface is under tension and the lower surface is being compressed (the reverse of a bridge, which is loaded in the middle rather than at the ends), but this was the era of the expansion of the railways in Britain, and he and his mathematician colleague were soon approached by rail pioneer Robert Stephenson to give advice on bridge building. In the design of the Conwy Railway Bridge (which is still standing) and the Britannia Bridge, both in Wales, Fairbairn and Hodgkinson took the box girder to its logical extreme and had the railway running through the middle of what they called a tubular girder.

A precast section of a box girder bridge is moved into position.

TRUSS BRIDGE

WHAT?

A type of beam bridge in which the load is carried by a wooden or metal framework that usually consists of triangular units.

WHERE?

Worldwide in a variety of applications from small road bridges to larger highway and rail crossings.

DIMENSIONS

Capable of a greater span than a simple beam bridge. The longest truss bridge span (that does not include elements of arch or cantilever design) is the 1,300-ft (400-m) Ikitsuki Bridge in Nagasaki Prefecture, Japan.

As we have seen, in a beam bridge the top of the beam is compressed and the bottom of the beam is under tension, or being stretched. The core of the beam experiences lesser forces. A beam can be made stronger by making it thicker, but this increases the weight and therefore the load on the middle of the beam. However, engineers discovered in the early 19th century that a beam could be effectively thickened and strengthened by adding lightweight truss work to the sides.

Usually made of steel, a bridge truss consists of two parallel "chords"—a top layer and a bottom layer—joined by vertical and diagonal bracing to form a network of triangles. A triangle is structurally very stable, the shape remaining constant providing the lengths of the sides do not change, unlike a rectangle, in which the sides can move relative to each other (which is why the joints in a rectangular structure are often braced with diagonal pieces that turn each corner into a small triangle).

The top and bottom chords of a truss take the forces of compression and tension respectively, while the joints, or nodes, between the members carry the load through the structure to the abutments at the two ends. This creates great rigidity, and in many architectural applications the truss takes the place of a beam. Trusses are also used to provide rigidity in

other forms of bridge design, such as arch (see p. 148) and cantilever bridges (see p. 146), as well as in roof supports.

First of the Modern Types

Light, strong, and relatively inexpensive, the truss bridge is one of the earliest forms of modern bridge, evolving initially in the UK in the first half of the 19th century against the background of readily available iron and the growth of the railways. Similar conditions in the USA led to a boom in this kind of bridge during the second half of that century, but here shorter-span road bridges were constructed using a combination of iron, which has great tensile strength, and timber, which has strength under compression. Many of the famous covered bridges found across the United States are made entirely from wood, and they showcase some of the main kinds of truss design used from the 1820s onward.

THE BAILEY BRIDGE

In the early 1940s, British bridge engineer Donald Bailey came up with the idea of a prefabricated modular truss bridge composed of 10-ft (3-m)-long panels that could be carried in a small truck, manhandled into position, and bolted together without the use of heavy machinery. The bridge design was adopted by the Corps of Royal Engineers and then by the Canadian and US forces, enabling the Allies to follow the retreating German and Italian armies despite the destruction of many bridges. By the end of the war, some 3,000 Bailey Bridges had been built, and many are still in use today. Donald Bailey was knighted in 1946 for his crucial contribution to the war effort.

VARIATIONS ON A THEME

The positions of the upright and diagonal members in a truss determine how the forces are transmitted through the structure, and throughout the 19th century engineers came up with a variety of designs, many of which were patented.

The King Post truss has a single central upright and two diagonals, but it is more commonly found as a roof support than in bridges. The Queen Post truss has two uprights, and this design is found in early covered bridges.

In 1840, William Howe of Massachusetts invented a truss in which the vertical members are under tension and the diagonals, which slope up toward the center, are being compressed. There are several covered bridges with this design, including the Jay Bridge in New York State.

Four years later, the Pratt brothers, Thomas and Caleb, came up with the inverse design, and this proved to work well for longer spans. Another four years later, James Warren's truss was patented, and this had no vertical members at all, just diagonals that support a varying combination of tension and compression as a load moves over the bridge.

In Australia, Percy Allan designed a truss that was widely used from 1890 onward, and there are several examples of his bridges still standing, including the Hampden Bridge in Wagga Wagga, NSW, which has cross-bracing connecting the tops of the two trusses.

CANTILEVER BRIDGE

WHAT?

A type of bridge in which the unsupported ends of two beams meet at the center of the span.

WHERE?

Commonly used for road and rail links, especially over waterways, as the construction method avoids disruption to navigation below.

DIMENSIONS

Although it can be used for short crossings and lightweight pedestrian use, cantilever design lends itself to extended spans, the world's longest being the 1,800-ft (550-m) span of the Quebec Bridge in Canada.

Cantilever bridges are usually made of trusses, which provide strength and rigidity with minimal weight, but sometimes of concrete and steel. Projecting beams are built out from each side of a channel, or equally out in both directions from piers constructed in the channel, without the need for costly scaffolding to provide temporary support during the construction process. These cantilevered beams then meet in the middle of the channel or support a central section between them.

A cantilever is defined as a member that projects beyond a fulcrum and is supported by a downward force behind the fulcrum or by a balancing member. An often cited example of the first case is a diving board, in which one end is bolted down and the other hangs out over the pool. This is known as an anchored cantilever. A balanced cantilever operates rather like a horizontal seesaw. The cantilever principle is also used in the construction of other types of bridge. For example, when a Bailey Bridge (see previous page) is being built, the bridge is gradually nosed out over the channel as more sections are added on the landward end, balancing the part that juts out until it reaches the other side.

First Cantilever Bridges

Some of the world's most famous bridges are cantilevers, and several have been in place since the 19th century. The very first cantilever bridge, built over the River Main in Germany and completed in 1867, was designed by the German engineer Heinrich Gerber. The first to be built in the USA was the Kentucky High Bridge, a rail bridge over the Kentucky River, designed by Charles Shaler Smith, a former engineering officer in the Confederate Army. Crossing a 275-ft (84-m)-deep gorge, when it was opened in 1877 it was the highest cantilever bridge in the world, and it held this honor for several decades. It was rebuilt and expanded in 1911.

Suspended Span

The Forth Rail Bridge, near Edinburgh on the east coast of Scotland, is the most famous of the early cantilever bridges. It was designed by civil engineers Sir John Fowler and Benjamin Baker, who were commissioned in 1883, and it came into operation in

QUEBEC BRIDGE

The world's largest cantilever bridge is the 1,801-ft (549-m) Quebec Bridge that crosses the St Lawrence River a few miles upstream of Quebec City in Canada. Work began in 1900, but construction was marred by tragedy. In June of 1907, as the work neared completion, workers found that some bolt holes in the lower chord of the southern cantilever arm were slightly misaligned. Engineers monitored the situation and found that the problem was worsening, but work continued.

Whereas in a truss bridge the top and bottom chords are under compression and tension respectively, in a cantilever, because it is supported at one end only, the opposite is true.

The deflection that the engineers were detecting was the bottom chords starting to buckle under the compressive forces. The bridge collapsed into the St Lawrence at 5:32 pm on August 29, killing 75 workers.

Construction on a redesigned bridge began in 1913, but on September 11, 1916, a further 13 construction workers lost their lives when the central suspended span plunged into the river as it was being lifted into position. The bridge was finally opened by the Prince of Wales (later King Edward VIII) in 1919, and to this day the central section of the Quebec Bridge remains the longest cantilevered span in the world.

1890. At the time, its gigantic 1,710-ft (521-m) steel truss spans were the world's largest, and it held this record until the completion of the Quebec Bridge some 17 years later.

The bridge consists of three towers, each 330 ft (100 m) tall, from which balanced cantilevers extend 680 ft (207 m) in each direction. On either side of the central tower, the cantilever ends support an additional 350-ft (107-m) suspended section bridging the middle of each span. This monument to civil engineering is made of some 50,000 tons (45,360 tonnes) of steel, including more than 4,000 tons (3,630 tonnes) of rivets!

The original High Bridge, Kentucky.

ARCH BRIDGE

WHAT?
A bridge with a curved form that carries the forces down to the ground through a support at each end.

WHERE?
Around the world, but China has the majority of the ten largest arch bridge spans.

DIMENSIONS
Arch bridges can have greater spans than beam bridges, and can even vie with cantilevers. Currently the longest span is that of the Chaotianmen Bridge in China, at 1,811 ft (552 m).

In this type of bridge, the deck is supported by an arch with its ends on either side of the channel or by multiple arches. The weight of the bridge itself and any load on it are carried outward and downward by the legs of the arch to the ground, where they exert a partially horizontal force through abutments. In a simple arch bridge, all the elements of the structure are under compression and there are no tensional forces, so the design lends itself to the use of materials, such as stone, that are strong under compression but have little tensile strength.

The concept of the arch bridge has been around for more than 3,000 years—the Arkadiko bridge in Greece dates from about 1300 BC—but it was the Romans who truly developed the idea. The majority of Roman bridges, of which many are still standing and still carrying traffic, had semicircular arches, although some were "segmental," the arch being composed of less than half a circle, giving a more flattened profile.

The Romans used stone, as well as an early form of concrete, and a bridge had to be supported by a timber framework, known nowadays as falsework, during the construction process. Once the central keystone was in place and the full weight of the bridge was allowed to act upon the arch, it was self-supporting. Indeed, with an incompressible material such as stone, the greater the weight upon

the bridge the stronger it becomes, which is why many stone bridges have a considerable depth of material between the top of the arch and the deck surface. The Romans soon discovered that the larger the arch, the greater the tendency for tensional forces to develop in the deck. Greater distances were therefore spanned by using a row of smaller arches, each with its own abutments. Some Roman multiple arch river crossings still exist. When greater height was required, for example to carry water across a valley (see p. 43), the Romans solved the problem by building rows of arches on top of each other.

Modern Developments
Over the centuries, many forms of stone and brick arch bridge have been built, including pointed and elliptical arches as well as open spandrels, a design in which there is a space between the arch, the deck,

and the vertical boundary of the bridge. However, with the development of new materials from the 18th century onward, many new possibilities were opened up. For example, in the 1770s the world's first cast iron arch bridge was built near Coalbrookdale in Shropshire, England. A monument to the Industrial Revolution, Iron Bridge is now a UNESCO World Heritage Site.

Since then, the use of steel trusswork and reinforced concrete have made it possible to rethink the position of the deck in relation to the arch—one doesn't have to be on top of the other!

In a through-arch bridge, the deck passes through the leg at each end of the arch and is suspended from it by cables or struts. The majority of the world's largest arch bridges are of this type. In a tied- or bowstring-arch bridge, which can look very similar, the deck itself is under tension, holding the two legs of the arch together and resisting the horizontal forces pushing them apart (a task normally performed by the abutments). In some respects, this type of bridge is not a true arch bridge but a variation on the truss bridge. New materials have also made it possible to create arch bridges that are less rigid than the fixed arch, by introducing hinges where the arch meets the abutment and also at the crown of the arch.

SYDNEY HARBOR BRIDGE

The steel through-arch bridge that has become a symbol not just of the city of Sydney but of Australia itself took a long time to come into being. The idea of building a bridge to link the northern and southern shores of Sydney's natural harbor was first put forward in the early 1800s, but it was not until after World War I that serious progress was made. The design for the bridge came from Dr J.J.C. Bradfield, who worked on the project from 1912 until its completion in 1932, and the Bradfield Highway, the main road that crosses the bridge, is named after him. The bridge was built by the English steel-producing and construction company of Dorman Lond, and the consulting engineer Ralph Freeman was knighted for his role in the project.

ARCH IN ACTION

Compression lines

In an arch bridge, the downward force of the load on the center of the arch is carried outward as a compressive force along the curve of the arch, resulting in both a vertical and a horizontal force on the abutments at each end.

SUSPENSION BRIDGE

WHAT?

A bridge, typically made of steel, positioned over a channel. Two long cables, anchored at either side, are "draped" over two tall towers, and the main bridge deck is suspended from these cables.

WHERE?

Usually over wide stretches of water.

DIMENSIONS

Variable, from a few tens of yards to 6,532 ft (1,991 m) (currently the longest suspension bridge span).

Suspension bridges are perhaps the most iconic of all engineering structures, noted for their elegance of form and transparency of function. The supporting towers rise up to three or four times the height of the horizontal platform (or "deck"), their verticality echoed by the suspenders that fall from the two main cables. Compared to other bridge types, the suspension is almost invisible. It is also one of the lightest of bridge forms, making them ideal for longer spans (heavier bridges carry the greater burden).

Suspension bridges have been around for a very long time, since early people first tied a vine rope across a space and pulled themselves across, hand over hand, dangling in mid-air. Later additions such as handrails and walkways eased the journey considerably! The more recognizable form of suspension bridge, with the deck suspended from cables strung above it, has

The two levels of the George Washington Bridge between Manhattan and New Jersey carry 14 lanes of traffic.

since evolved through better technical understanding and quality of materials. One of the first of the modern type of suspension bridge, completed in 1826, was built over the Menai Straits in Wales by Thomas Telford. Many famous suspension bridges have been erected since then, including the Brooklyn Bridge, the Golden Gate Bridge (see box), and the Akashi-Kaikyo Bridge in Japan, currently the world's longest. Such projects have often flown in the face of received wisdom by bridging what had previously been seen as unbridgeable. The suspension model is highly adaptable and can be constructed with as much complexity or simplicity as required. The Bombi Bridge, a 366-ft (111-m) footbridge over the Galana River in Kenya, was constructed in 2009 by local people, aided by Western engineers, using only hand tools. A charitable project, it stands as a fantastic example of the power of civil engineering to change people's lives.

Construction

The construction of a wire strand cable-type suspension bridge can take anything from 18 months to ten years. First up are the towers, as well as the two huge anchorages at either side. These act like tent pegs, securing each of the bridge's main cables. They are driven deep into either concrete or bedrock (though occasionally into the end of the road deck). A pilot line following the projected trajectory of each cable over the two towers is then installed, along which wire is then "spun." The wire is secured to one of the anchorages, then spun back the same way to the other side. The process is repeated until a strong enough bundle of wires is formed and fused.

Once sufficient bundles have been spun, they are compressed hydraulically and coated to form the main, massively powerful cable. Vertical suspender cables are then placed at regular intervals along each of the two main cables. Prefabricated deck sections are lifted into place by barge or derrick and are attached to the suspenders. Deck construction must proceed in both directions from the support towers at an even pace to maintain balance. Once completed, the bridge is finished with handrails, paint, lighting, and so on.

THE GOLDEN GATE BRIDGE

Completed in 1936, the Golden Gate Bridge is the oldest of the world's top ten longest suspension bridges, and probably the best known. Linking San Francisco to Marin County in the north, its main span reaches 4,200 ft (1,280 m). Some 80,000 miles (128,748 km) of wire form each of the main cables. Although its name suggests yellow, and it looks red, the Golden Gate is actually painted International Orange, a standardized color often used in engineering projects to set structures apart from their locale. A symbol of San Francisco, California, and the USA, the bridge also promised a new life to thousands of Asian immigrants in the early 20th century. Ironic, then, that it has since become one of the world's most popular places to commit suicide.

SUSPENSION BRIDGE ANATOMY

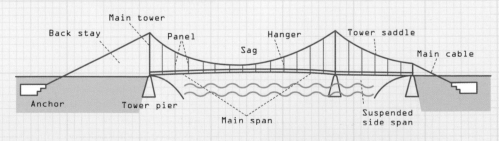

In a suspension bridge the deck is literally suspended from hangers attached to main cables that run the full length of the bridge and pass over supporting towers. The load translates into a tensional force on these cables, which must be securely anchored at each end.

CABLE-STAYED BRIDGE

WHAT?

A bridge, usually a continuous girder, supported by cables that run from the deck directly to supporting towers.

WHERE?

Usually over wide stretches of water.

DIMENSIONS

Suitable for spans in excess of those that can be bridged by cantilever methods, cable-stayed bridges are more cost-effective than suspension bridges in certain circumstances.

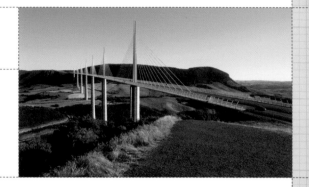

Although the appearance of a cable-stayed bridge is in many ways similar to that of a suspension bridge, and the deck is indeed suspended from cables, the design principles are quite different. Instead of the load being carried by main cables that run the length of the bridge, the deck, which is usually a continuous girder, is supported by cables that run diagonally to one or more tall towers positioned along the length of the bridge.

The Brooklyn Bridge borrows from the cable-stayed design.

The towers of a cable-stayed bridge transfer the weight of the deck down to the ground, whereas much of the weight of a suspension bridge is borne by the anchorage points of the main cables. This gives the cable-stayed bridge an advantage where soil conditions make it difficult to secure a good anchorage, but the towers need to be built on stable and solid piers. Like a cantilever bridge, a cable-stayed bridge can be built out from the supporting towers, with bridge sections and cables gradually being added equally on both sides of each tower.

From Then Till Now

A cable-stayed bridge, often with slender masts and barely visible cables and sometimes having a curved or arched deck, can appear to hover gracefully above the water, but despite its modern look this concept has been around for a long time. In a book entitled *Machinae Novae*, published in Venice in 1595, the Croatian polymath Faust Vrancic revealed his plans for a bridge held up by chains attached to two towers, but it was not until the second half of the 19th century that such bridges were actually built and even then they tended to be hybrid designs that relied only partly on cable support. For example, New York's Brooklyn Bridge, opened in 1883, is a suspension bridge, but it also has cables running diagonally from the towers to the deck to give it added rigidity.

Cable-stayed bridges of the kind we see today first began to appear in the 1950s, and there are now many beautiful examples around the world. Simple pairs of upright towers have given way to X- and Y-shaped towers, single towers, backward-sloping towers, and wishbone towers, and in place of straight, flat decks, arched and curving highways sweep through the air. The spans of these bridges has steadily increased, too, and China now boasts two cable-stayed bridges with spans in excess of 3,300 ft (1 km), the largest being the Sutong Bridge across the Yangtze River, with towers more than 1,000 ft (300 m) tall.

Cradle Stay

With the supporting cables anchored to the sides of the tower, there is a significant force pulling the tower apart, and the structure has to be strong enough—and thick enough—to resist this. In a recent innovation, the cradle-stay system, continuous cables run from the deck on one side of the tower,

BRIDGE ANATOMY

The cables that support the deck are arranged in two main ways. In a "fan" design, the cable attachments are grouped at the top of the tower and the cables fan out to meet the deck along its length. In a "harp" design, the cable attachments are spaced out on the tower and the cables run parallel with each other down to the bridge deck.

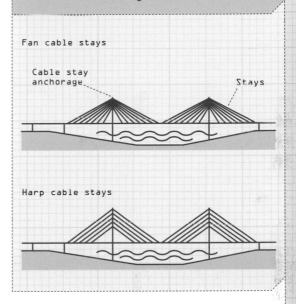

Fan cable stays

Cable stay anchorage

Stays

Harp cable stays

through a curved tube, or cradle, within the tower, and down to the deck on the other side. As the towers need to resist only a downward, compressive force, they can be much slimmer, making construction faster and cheaper. Each cable is made up of many separate strands running parallel to each other, so these can be inspected and replaced as necessary one by one without compromising the bridge's integrity. The system, designed by Figg Engineering Group, was first used on the Toledo Skyway Bridge in Toledo, Ohio, and then in the construction of the Penobscot Narrows Bridge near Buckport, Maine, which features an observatory atop one of the two 420-ft (128-m)-tall towers.

The Penobscot Narrows Bridge uses the cradle-stay system.

MOVABLE BRIDGE

WHAT?

Bridges that can be lifted, swung aside, or withdrawn, generally in order to allow the passage of shipping on a waterway.

WHERE?

Across navigable waterways worldwide.

DIMENSIONS

The maximum length depends on the type of bridge, but single spans rarely exceed 250 ft (76 m).

In many situations, a high-level bridge can be built to keep road traffic clear of shipping, but sometimes the high cost of construction or a lack of space to create the change in level at each end prohibits this solution. This is where the movable bridge has a role. The methods of construction used in these bridges vary, although most have a girder or truss structure, but they all share the ability to move upward or sideways to allow tall vessels to move through the channel. When in their closed positions, movable bridges carry pedestrian, rail, and/or road traffic.

SINGLE LEAF BASCULE BRIDGE

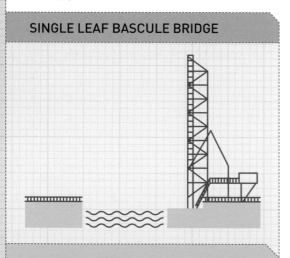

The leaf of a bascule bridge is designed to be slightly heavier than the counterbalancing weight so that it has no natural tendency to rise. The motor must be powerful enough to lift the leaf and any water or snow that may be on it.

Over the centuries, a wide range of bridge designs, including some harebrained schemes, has been proposed to solve the competing needs of highway and waterway traffic, but the designs in use today boil down to three main types. Each has its merits and advantages in particular situations, and the choice of which to use depends on such factors as the length of the span, the width of the carriageway, the load that will be carried, and the frequency with which the bridge has to open.

Bascule Bridges

The bascule bridge is the most common form of movable bridge. The term bascule is the French word for a seesaw (the word is actually a contraction of the French for "bump-buttocks!"), and in this design the bridge deck is counterbalanced by a weight on the landward side. The whole assembly can then be pivoted about a fulcrum close to the balance point with relatively little effort—a small electric motor is enough to raise the bridge, making this design relatively quick and inexpensive to operate. Usually the counterweight is hidden beneath the roadway, but it can take the form of blocks on either side of the road or a single weight held aloft above the road itself. On wider spans, the bridge has two parts or "leaves"—one on each side of the channel. London's Tower Bridge (pictured on the opposite page) is a classic example.

Government Bridge swings to allow shipping through.

Swing Bridges

The typical swing bridge is a truss bridge mounted on a central pier. A central bearing allows it to rotate and align itself with the channel, allowing ships to pass on both sides of the pier. This design is cumbersome and slow to operate, but it has all the qualities of a cantilever truss bridge and can span wide waterways. Many swing bridges carry railroads and date back to the end of the nineteenth century. The Government Bridge, for example, that crosses the Mississippi River at Rock Island, Illinois, was built in 1893, spans a 500-ft (150-m) channel, and carries road and rail traffic on two separate decks.

Vertical Lift Bridges

As the name suggests, this kind of bridge is raised in its entirety. The movable section is suspended from counterweighted cables that pass around rotating drums in a tower at each end, and these are electrically operated to lift the bridge like an elevator. This kind of bridge can be extremely sturdy, making it suitable for rail traffic, but the clearance it creates is limited by the height of the towers at each end.

ODD ONE OUT

One of the strangest movable bridges is the Rolling Bridge on London's Grand Junction Canal, though it doesn't so much roll as curl (see the inset photograph above). This footbridge is composed of eight triangular steel sections, the tops of the triangles on each side of the deck being joined by hinged links. As hydraulic rams between the deck and the links push the hinges upward, the tops of the triangles are pulled together and the bridge curls back to form an octagon on one side of the channel. The reverse happens when the rams pull the hinges down again and the bridge extends out over the water.

Buzzards Bay Railroad Bridge spans the Cape Cod Canal.

PORT

WHAT?

A facility located on a coast, river, or lake where boats and ships can dock to load and unload passengers or cargo.

WHERE?

Every country that has a coastline—and even some that don't—has a port of some kind.

DIMENSIONS

The "size" of a port is calculated on criteria such as the ship tonnage, number of containers, or overall weight of freight handled. Shanghai is currently the leader in terms of total cargo, handling some 650 million tons (590 million tonnes) in 2009. The total length of its quays is 12.4 miles (20 km).

Ports are a crucial link in the global transport network for natural resources, materials, and goods, and many of the world's major cities owe their growth and prosperity to their importance as ports—Mumbai, New York, Manila, Rotterdam, London, Shanghai, Hamburg, Buenos Aires, Houston, Antwerp, and Vancouver to name but a few. Many are located on major rivers.

Although the terms port and harbor are often used synonymously, a harbor is a protected natural or artificial haven for boats, whereas a port, while offering shelter to shipping, is also equipped with facilities for embarking and disembarking passengers and/or for loading and unloading, storing, and transporting cargo. There are several distinct types of port, although large ports may fulfill all of these functions.

Port Basics

Small ports can be located in natural harbors, but larger seaports have artificial concrete arms or breakwaters that project out at an angle from the coast and protect the enclosed space from wind and waves. Major ports in the naturally protected waters of a river or inlet, such as the Port of Vancouver, can be spread out over a large area, with separate wharves distributed along the shoreline.

The width of the entrance, the depth of the channels, and the turning space within the port dictate the maximum size of ships that can use the facility. Deep-water ports can handle the largest and most lucrative ships, but many ports have dredgers to maintain the necessary depth. There will also be marine services such as marine fueling, tug and barge towing operations, handling and transfer of ship's waste, and facilities for maintaining and repairing ships, often including some form of dry dock. A floating dry dock is a structure with pontoons that can be filled with water to submerge it so that a ship can enter. The pontoons are then filled with air and the structure raises the ship clear of the water. Some ports have a "graving" dock, a narrow channel that can be drained to leave a ship resting on blocks. Cargo ports are almost always transport hubs, providing links to railroads, highways, and sometimes canals by which goods and materials arrive and depart. Many ports also have warehouses and cold-storage facilities, as well as industries for processing, packaging, and distributing the goods that pass through it.

Fishing Port

Many ports, especially those that are focused around a natural harbor, began life as fishing ports. The primary function of a modern fishing port is to provide amenities for commercial fishing vessels such as large-scale trawlers. These include berthing, offloading for the catches, transport links, and sometimes storage and processing facilities, as well as repair workshops and chandlery supplies. Many fishing ports also cater for recreational anglers and boaters, offering fishing charters and small boat marina facilities.

Some small fishing ports have been in existence for hundreds of years, or even thousands in the case of the Mediterranean coast, but the modern commercial fishing industry is highly dependent on the market, on the state of fish stocks, and on regulations, all of which can cause the fortunes of particular fishing ports to change over time, some expanding and others losing their fleets altogether.

Passenger Port

Cruise ship terminals consist primarily of berthing facilities for large passenger liners (including a range of vessel services such as fresh water provision and garbage collection), and amenities for passengers embarking and disembarking, such as customs services, check-in, and connections to road, rail, and air links (airport, heliport, float plane terminal). Car ferry terminals require more infrastructure and take up far more space, as they need car parking facilities for foot passengers as well as toll booths and holding areas for vehicles waiting to board.

Bulk Cargo Port

Dry and liquid bulk cargoes consist of materials and substances that are not packaged but are transported in bulk and can be transferred to and from cargo ships using pumps and pipes, bucket elevators, and conveyor belts. Many ports grew up around the

Port Newark-Elizabeth Marine Terminal, part of the Port of New York and New Jersey, is a major container freight facility.

shipping of one particular resource, such as coal, iron ore, or grain, and some still specialize, but most major ports have facilities for handling a range of bulk materials. These facilities, or terminals, are usually located in separate wharves within the port and are owned by individual companies that export or import particular goods, such as fuel oil, wheat, sulfur, wood chips, or raw sugar. There are also multi-product bulk terminals that can handle anything from coal and animal feed to fertilizers and vegetable oil. Bulk terminals are equipped with loading and unloading equipment, storage facilities, and road and rail links.

Containerization

Packing nonbulk cargo in standardized containers has revolutionized the transport of freight. Almost 90 percent of all non-bulk cargo is now moved in containers, and the heart of many ports is the container terminal, a facility for loading and unloading containerized cargo to and from ships. From the beginning of ship transport, cargo that could not be moved in bulk was "break bulk cargo"—boxes, sacks, cartons, drums, barrels, reels, and pallets of every size and description that had to be loaded and unloaded as separate items. The process was slow, labor-intensive, and prone to theft and breakage, and there had been attempts to rationalize the system, but none had caught on. Standard boxes that could be transferred from barges to railcars were used for the bulk transport of coal in Britain and Germany in the 18th century, and in the USA in the first half of the 20th century railroad boxcars were carried by ships and road-going trailers

TYPICAL CARGO PORT LAYOUT

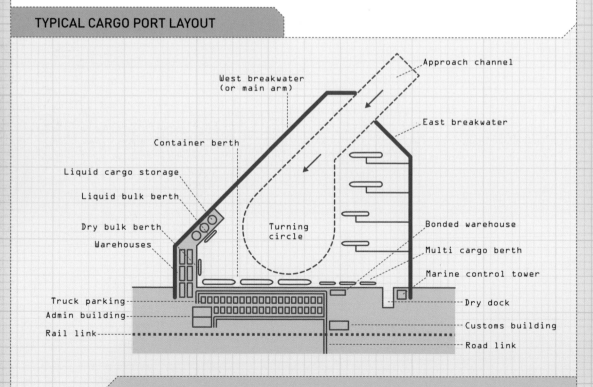

Every port is different, and the layout will depend upon factors such as the form of the coastline, the depth of the water, the scale of the operation, and the types of cargo that dominate the port's use, but most ports that handle general cargo will have many of the features shown here.

were carried on railroad flatcars, but it wasn't until the 1950s that the standardized "intermodal" shipping container came into being.

It took a while to reach international agreement on all aspects of it, but the modern container is a strong steel or aluminum box with standard castings on the corners that allow it to be lifted by specialized equipment, secured in a ship, on a railcar or highway trailer, and stacked with other containers. The length of a container is generally a multiple of 10 ft (3 m), and the 20-ft (6-m) equivalent unit, or TEU, is used as the unit of measurement for a vessel's container capacity and for the amount of container freight passing through a port (an FEU is two TEUs). There are also 45-, 48-, and 53-ft (13.7-, 14.6-, and 16-m) containers. The standard container width is 8 ft (2.4 m) or 8.5 ft (2.6 m), and the height is 8.5 ft or 9.5 ft (2.9 m). Within these parameters a container may be open-topped, refrigerated, designed for dry or liquid bulk cargo, or otherwise specialized in a number of ways.

Container Terminal

Containerization has many advantages. Containers can be sealed and locked, keeping the contents safe from theft, and sealed containers can be more quickly moved through customs, as the contents generally do not need to be examined. Their "intermodal" quality means that containers can be quickly and easily transferred between purpose-built ships and highway transport trailers or railroad cars that are specifically designed to carry containers. This transfer takes place at the port's container terminal using specially designed lifting equipment. The container ships themselves are huge—some are able to carry more than 14,000 TEUs—and ports that handle container freight are often correspondingly large. The port of Singapore handled almost 26 million TEUs in 2009. Although the majority of containers unloaded from ships continue their journey inland, many container terminals also have a container freight station where cargo is taken from containers for distribution, and goods are "stuffed" into containers for loading.

THE GANTRY CRANE

Giant cranes and forklift machinery are used throughout ports for loading and unloading all kinds of cargo, but the gantry crane is the key to smooth container movement. It consists of a tall frame (gantry cranes can weigh over 1,000 tons [900 tonnes]) that runs on rails and has a hoist mounted on a trolley that can be sent out over the container ship on a long boom. A spreader bar at the end of the hoist cable locks onto the corner castings of the container, which is then raised and brought through between the legs of the crane to be stacked or loaded onto a railcar or truck. Mobile cranes and top-lifting container handlers are used for moving containers around within the port.

COASTAL DEFENSES

WHAT?
Concrete, stone, or timber structures designed to defeat the scouring and eroding action of the sea.

WHERE?
Wherever the action of the sea has a negative impact on important land or structures and the economic resources are available to prevent it.

DIMENSIONS
The world's longest breakwater is the 6.74-mile (10.85-km) granite South Breakwater at the Port of Galveston in Texas. The longest sea wall is the 21-mile (34-km) Saemangeum sea wall in South Korea.

The coastline of every island and every continent is continuously subjected to the action of the waves, winds, and tides. These forces constantly sculpt the shoreline, eroding the rock and carrying away sand and pebbles. On a global scale the process is unstoppable, but where shipping, housing, agricultural land, or recreational amenities are threatened, engineering solutions are put in place to slow the process down.

We have already seen how concrete arms are used at the mouth of a harbor or port to break the force of the wind and waves and protect the boats and ships within. If they are correctly positioned, they can also have the effect of preventing silt from building up in the entrance channel, and these two functions—protection and current control—are put to work along coastlines in a number of ways.

Groins
Many coastlines experience a phenomenon known as longshore drift, or littoral drift. This occurs where the prevailing wind is parallel, or at a shallow angle, to the shore, causing waves to wash up the beach at an angle before running back to the sea. The net effect is to move small particles—mainly sand, but also shingle and small pebbles—along the shore in the direction of the wind.

In many resort areas and locations where this kind of erosion is a serious problem, structures called

groins (or groynes) are used to absorb the energy of the waves and reduce movement along the shore. Groins are rigid structures, built of rock, concrete, or heavy wooden boards fixed to timber piles set into the beach, that extend out into the water at right angles to the shore. By preventing movement along the beach, each groin maintains the sediment on its "updrift" side and can actually help to build up the beach. In an effect called terminal groin syndrome, the shore on the downdrift side of the last groin in a series can become severely eroded as it is unprotected and starved of new material.

Concrete groins can also be used as "entrance training walls" to keep a channel deep and free from silt where a river flows out between sandy beaches, although by blocking the natural process of longshore drift the groins may have the effect of increasing erosion on either side.

Sea Wall

In areas of low-lying coastal land, sea walls perform the vital role of not only preventing erosion but actually keeping the sea out. Historically, sea walls had a vertical face that met the power of the waves head on, but this caused damage to the wall itself and was found to enhance longshore drift. Sloping walls with a recurved top edge to turn the wave back and absorb its energy are now more common.

Revetment

This is a shallow sloping wall that absorbs the power of the waves more gradually without affecting the passage of sediment and currents along the shore. Revetments are typically made of specially shaped flat-faced stone or of riprap, a mixture of irregular boulders, the uppermost of which may weigh several tons. In some situations, revetments are constructed of timber piles and sloping wooden planks, somewhat like a groin but running parallel to the coast.

Riprap Gabion

This unlikely term refers to cubes or long sections made by filling wire mesh cages with rough rock to form a wall. Although gabions do not last as long as concrete structures, they have the advantage of allowing the sea to flow through them to some degree, and they trap sand and stones between the rocks to form a strong defense.

THE TETRAPOD

These four-footed cast concrete structures were invented in France in the early 1950s and are now used—together with various multilegged variations—around the world to build harbor arms, groins, and protective revetments. Reviled by many as ugly monstrosities and a hazard to swimmers, they are nonetheless well qualified for the job. They last for a long time, their shape allows the water to flow round them, dissipating the energy of the waves, and their legs interlock, making them difficult to dislodge.

RECURVED SEA WALL

Although they are expensive and can interfere with public beach access, seawalls have a long lifespan and are highly effective in preventing the erosion of cliffs and coastal land. The concave face turns the wave back on itself rather than receiving the full force of the impact.

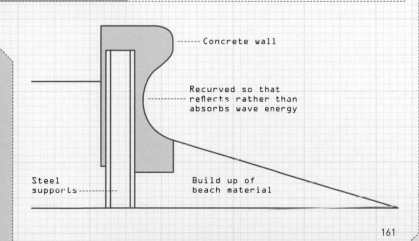

Concrete wall

Recurved so that reflects rather than absorbs wave energy

Steel supports

Build up of beach material

LIGHTHOUSE

WHAT?

A tall, lighted aid to navigation positioned in or near navigable water to provide visual guidance to mariners.

WHERE?

Wherever ships and hidden rocks are likely to meet. There are estimated to be more than 12,000 lighthouses dotted around the globe, all of them marking hazards to shipping.

DIMENSIONS

The top of a lighthouse must be tall enough for the beacon to be seen from a distance On a high cliff, a lighthouse need only be 20 ft (6 m) tall, but many built at sea level are more than ten times that height.

Few functional structures elicit as emotional a response as the lighthouse, a symbol of isolation, heroism, and the battle against the forces of nature. The number of lives saved by lighthouses is incalculable, and many, perched on wave-battered rocks, are marvels of engineering. Sadly, the days of the lighthouse may be numbered—unless lighthouse lovers have their way.

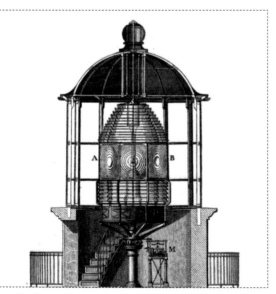

The Fresnel lens made lighthouse beams much brighter.

The lighthouse has a long history, dating back to at least the third century BCE, when the Lighthouse of Alexandria was built on the island of Pharos on the coast of Egypt to mark the entrance to the harbor. Possibly as much as 450 ft (140 m) tall, it was one of the Seven Wonders of the World until, after suffering damage from a series of earthquakes in the 12th and 14th centuries, it was finally demolished in 1480. The word "pharos" is the root of the word for lighthouse in several languages, including French (*phare*).

The Chinese built lighthouses to mark the entrances to ports, and the Romans used bonfire-topped towers as markers for shipping. At Dover, on the south coast of England, one of two such towers built in the first century CE is still standing, and there are several on the shores of the Mediterranean, but most of these fell into disuse. The oldest lighthouse still in use is the Tower of Hercules, built by the Romans more

GOODBYE LIGHTHOUSE KEEPER

When lighthouses had oil-burning wicks and clockwork mechanisms to turn the lens they needed constant supervision, and every lighthouse was manned—even the most remote and isolated. However, the development of technology such as solar power, remote monitoring, and automatic bulb changing has led to a steady process of lighthouse automation over the last few decades. The USA, for example, no longer has any permanently staffed lighthouses. In some countries, such as Canada (where there are still some 50 manned lighthouses), the move toward destaffing is being fought on the grounds that lightkeepers also provide an important rescue service and vital up-to-the-minute weather information.

than 1,800 years ago on a peninsula at La Coruña on the northwest coast of Spain. The area is known as the Coast of Death (Costa da Morte) because of the large number of ships that have been wrecked here. From the 15th century on, as shipping and maritime safety became increasingly vital for trade and naval activity, an increasing number of lighthouses were built in northwest Europe, and the great age of lighthouse building reached its peak in the 18th and 19th centuries, especially in Britain and the USA.

Lighthouses around the world range from light steel or wooden structures to imposing monuments, but one of the most iconic and enduring is the Fastnet Lighthouse on the Fastnet Rock, the most southerly point of the Republic of Ireland. Completed in 1904, the 146-ft (44.5-m) tower is built of massive interlocking granite blocks and is topped by an electric light that can be seen 27 nautical miles (50 km) away. For more than 100 years, the lighthouse has withstood the fury of the North Atlantic storms, including the impact in 1985 of a rogue wave that was the same height as the light.

Lighting the Beacon

Early lighthouses had a bonfire, candles, or an oil-burning wick to signal the presence of the coast, a reef, rock, or shoal. In the 19th century, the light was made to flash—each lighthouse had its own distinctive pattern of flashes—by using a rotating curved mirror or thick lens to concentrate the light into a beam that swept round in a circle, but the system was greatly improved by the introduction of the Fresnel lens. Composed of multiple prisms, it is much lighter than a conventional lens and focuses more of the light into a tall, thin beam. By the beginning of the 20th century, most lighthouses were equipped with Fresnel lenses (see opposite) and powered by electricity.

The development of GPS and other forms of navigation system has cast a shadow over the whole future of lighthouses, and many have already been shut down. However, in some parts of the world communities and charitable organizations are stepping up to fight for their preservation, and even to purchase and maintain them.

AIRPORT TERMINAL

WHAT?
A complex of buildings designed to accommodate passengers departing and arriving by airplane.

WHERE?
Close to every major city in the world. There are almost 900 airports worldwide that handle international flights, and there are many very large domestic airport terminals.

DIMENSIONS
The world's largest terminal, Terminal 3 at Dubai International Airport, covers an area of 370 acres (1.5 km²).

There was a time when an airport terminal was a one-room building in which passengers waited to board a small plane on a single strip of concrete outside. Now that planes can seat 800 people, the runway may be 3 miles (5 km) long, and the airport terminal is a huge complex with all the facilities of a small city.

This dramatic change in scale of the airport terminal is a consequence of the sheer volume of air traffic, which in a hundred years has gone from zero to over 77 million flights a year carrying more than 4.88 billion passengers (in 2008—international flights have decreased slightly over the last few years). In 2010, the world's ten busiest airports handled an average of almost 60 million passengers each, so they have to be big—and organized.

Getting There
Large airports are always located close to cities, and are connected to them by a range of transport systems—major highways, railroads, possibly subways or light rail links—and terminals have the stations, road systems, and extensive car parks that these require, as well as offering landing facilities for small aircraft and helicopters. Car rental companies have their own office space and car parking areas at every major airport. In airports that comprise several terminals, connections between them are made by automated rapid transit systems.

Checking Through
The internal layout of the terminal is designed to lead passengers through the various steps in getting to their plane as smoothly as possible, initially from ticket sales and baggage check-in, through customs and security checks from "land-side" into the "air-side" part of the terminal. Wherever they are in the terminal complex, passengers are kept advised of departure times and gate numbers by means of Flight Information Display (FID) systems, and while they wait for their flight to be announced the airport provides them with every amenity, from washrooms and Internet to restaurants, banking facilities, and dozens of stores. When it is time to head for the departure gate, which can be some distance from the main area, there are moving sidewalks and courtesy vehicles to help.

Behind the Scenes

Passengers see only a fraction of what is going on at the airport. Beyond the public space there is an entire service industry at work, delivering goods, food, and equipment to the airport, maintaining, cleaning, and refueling the aircraft, supplying in-flight meals, controlling the movement of airplanes on the ground, organizing the loading and unloading of cargo and mail—the list is endless. And on top of this, all the many workers (70,000 people work at London's Heathrow Airport) require their own services and amenities. The airport really is a small city.

Bridging the Gap

Parking the planes close to the terminal building means that passengers don't have to walk far on the airport apron, but they are still out in all weather, still able to wander off and get into trouble, and they still have to climb up steps to plane doors that are getting ever higher from the ground. All that changed in the 1960s with the introduction of the jet bridge, a maneuverable telescoping corridor that connects to the departure gate of the terminal at one end and to the airplane doorway at the other, allowing passengers to walk from one to the other along a gentle slope—secure, warm, and dry, and often without even realizing they are leaving the terminal.

The jet bridge is a work of engineering in its own right. At the terminal end, it may be permanently attached to the departure gateway with a rotunda, an articulated circular structure that allows the bridge to pivot, or there may be a fixed walkway between the gate and the rotunda.

The bridge itself may have permanent fixed supports along its length or it may be supported by a wheeled frame that allows it to swing sideways. The airplane end of the jet bridge is equipped with controls enabling the operator to raise, lower, and turn the end of the bridge to match up with the position of the airplane doorway. The corridor can also be extended and retracted.

LUGGAGE HANDLING

One of the key engineering components of an airport terminal is its baggage-handling system, running invisibly beneath the concourse and ensuring that every bag makes its way from the check-in to the right plane and from one plane to another. The whole system is automated and computer-controlled, with scanners that continually track the bar-coded baggage labels and mechanical arms that push and divert individual bags through a network of conveyor belts and chutes. The system can detect bag jams and can even route baggage around them.

In some airports, separate jet bridges can be attached to the front and rear doors of the plane to reduce boarding and exit times, but in the last few years duel-end bridges have been built that connect both doors to one gate by branching and overflying the airplane wing.

Jet bridges are in use in all major airports.

TERMINAL LAYOUT

The primary purpose of the terminal is to connect people and flights. At one time (and still in very small airports) passengers simply walked or took a bus across the apron (the airplane parking area) to the waiting plane, but this doesn't work on the scale of a large international airport and it is necessary to get the planes as close as possible to the passengers. The overall layout of the terminal is designed to achieve this by maximizing the outside perimeter of the terminal and providing airplane parking spaces around it. Every terminal is different, but there are some fundamental concepts on which most are based.

Linear: A terminal can simply get longer, providing more parking on the runway side and increasing the interior space for passenger "processing."

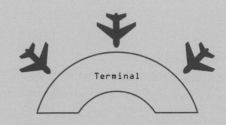

Semicircular: A horseshoe shape provides even more parking spaces for the same floor area, and many airports use this design.

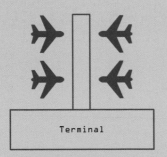

Pier: The concept of adding piers that stick out from either a linear or a curved building has been around since the 1950s. Planes can be parked at departure gates along both sides of each pier. The downside is that passengers have a long trek from check-in to departure gate, necessitating the introduction of travelators and even monorails.

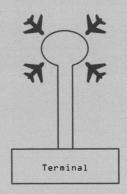

Pier Satellite: Expanding the tip of the pier into a circular building enables more airplanes to cluster around it, as well as along the length of the pier.

TERMINAL LAYOUT [CONT'D]

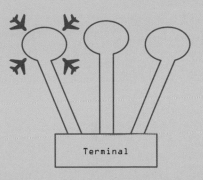

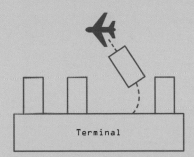

Remote Satellite. Instances of this design are relatively rare, but here the pier is dispensed with and the passengers make their way from the terminal to the departure gates by means of an underground walkway or subway system.

Mobile Lounge: This concept has a lot in common with taking a bus to the plane, but in this case the bus is the departure lounge itself, in some cases mounted on scissor jacks to raise the carriage up to the height of the airplane doors. Although they allow airplanes to park further from the terminal, very few are still in use, largely because they have been superseded by the jet bridge (see below).

JET BRIDGE

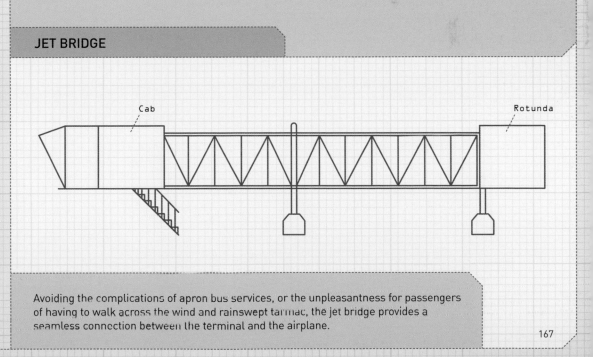

Avoiding the complications of apron bus services, or the unpleasantness for passengers of having to walk across the wind and rainswept tarmac, the jet bridge provides a seamless connection between the terminal and the airplane.

167

AIRPORT RUNWAY

WHAT?
Flat paved strip on which airplanes land and take off.

WHERE?
Every airport has at least one runway, and some have many more. Dallas/Fort Worth International Airport has the most, with seven.

DIMENSIONS
The largest international wide-body jets require a runway of at least 13,000 ft (4 km), and a few airports have runways 16,400 ft (5 km) long. The minimum width for large commercial passenger jets is 150 ft (45 m).

Pilots, ground crew, and air traffic controllers around the world need to be on the same page when it comes to dealing with runways and there is, of course, an internationally agreed and accepted set of regulations and practices. These cover everything from the runway numbering system and distance indicators to airport and runway lights. Even as a passenger, it is possible to pick out many of these features as the aircraft comes in to land, and the lighting system can be seen clearly on a night flight.

A runway at a major airport is much more than just a very long strip of road in the middle of a very large field. The orientation of the runway and its position relative to other runways need careful planning, and every runway must have an identification number, not to mention access taxiways, a host of painted symbols and markers, and color-coded lighting to guide and warn pilots and ground crew.

Runway Layout
In the early days of aviation, "flying fields" with no runways made it possible for aircraft always to take off and land into the wind, shortening takeoff and landing distances and avoiding crosswinds. When a modern runway is being built, the same consideration applies, and the prevailing wind (the direction from which the wind most commonly blows) is taken into account when determining the orientation of the runway. There is also the noise issue to consider and designs try to avoid, where possible, sending low-flying aircraft over populated residential areas. The existence of other runways is also a major factor, and airports try to avoid complex intersections between runways and taxiways, or placing two parallel runways so close that a plane taxiing to or from one affects the other. If possible, airport vehicles are routed along separate perimeter and service roads, or even tunnels, so they don't have to cross runways.

There are three basic layouts for positioning two runways, and most airports have a variation on, or a combination of, these:

Parallel runways obviously run parallel to each other, and must be separated by a certain distance. If one runway is used only by light aircraft, the separation may be less than 2,500 ft (760 m), but if both are used by large jet aircraft they will be at least 4,500 ft (1,370 m) apart.

In an open-V layout (see right), the two runways diverge from each other but don't intersect, forming a V with the point cut off. In strong wind conditions, this layout offers a choice of which runway to use, and when there is little wind both can be used. This layout can handle a greater number of takeoffs when planes are taking off from the "bottom" of the V than when they are converging on takeoff.

Intersecting runways cross each other and, because they run in very different directions, can handle strong winds from many directions. In light wind conditions both can be used, but the timing of landings and takeoffs becomes a vital safety issue.

Naming Runways

In the words of the US Federal Aviation Administration, "Inadvertent entry onto an active runway represents one of the most persistent hazards in civil aviation." It is essential that pilots and airport staff identify individual runways correctly, from the air and on the ground, and therefore an international system for naming and marking them is used. Runways are numbered according to the compass direction in which they run.

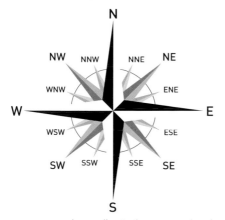

Runways are named according to the compass bearing in which they point.

RUNWAY LAYOUT

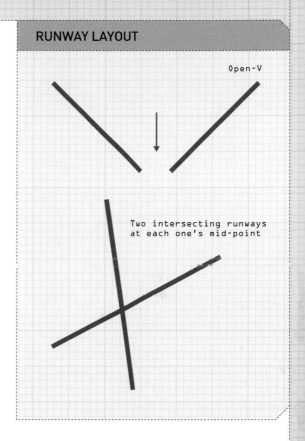

Open-V

Two intersecting runways at each one's mid-point

North is zero/360°, east is 90°, south is 180°, and west is 270°, and to simplify matters, the compass heading is rounded to the nearest 10 and the final 0 is removed. So a runway that has a heading of 107° will be runway 11, and this number is painted at the start of the runway. If it were the right-hand one of two parallel runways, it would be 11R. The center one of three would be 11C.

Of course, every runway can be used in both directions, so from the other end this same runway, which now has a heading of 287° (107°+180°) is runway 29, and the runway as a whole is referred to as Runway 11/29.

Australia's Sydney International Airport, for example, has two parallel runways called 16R/34L and 16L/34R, one of which is intersected by Runway 07/25.

Runway Markings

The number, or designation, of the runway is only one of the many pieces of information painted on the tarmac. Areas before and beyond the runway proper may be marked with yellow chevrons indicating overrun areas that are not meant to be used for taxiing or takeoff. There may then be a section of runway that is good enough for taxiing but not for landing or takeoff, indicated by long white stripes. The threshold—the true start of the runway—is marked by two blocks of parallel longitudinal white lines followed by the runway number. A striped centerline begins at this point and continues the length of the runway.

The next thing the pilot sees (depending on country) is one, two, or three long white lines on each side of the centerline, marking the start of the touch-down zone. This is followed 300 meters later by a fixed distance marker—a block of thin white longitudinal lines indicating the ideal touch-down point—and then an "end of touch-down zone" marker. Some runways have, instead, an aiming-point marker consisting of a broad white rectangle on each side of the centerline. This is aligned with the VASI or PAPI lights beside the runway (see box opposite).

Taxiways also have markings, but these are in yellow. The most important of them is the "runway holding position marking," two dotted yellow lines across the taxiway preceded by two solid yellow lines. This is the last stopping position before entering a runway, and the pilot must check that the runway is clear before continuing.

Runway Lighting

To guide pilots in their approach, landing, taxiing, and takeoff at night and in poor visibility, signal lights are used. At civilian airports, to help pilots distinguish between runways and highways, a highly visible beacon on the airdrome flashes green and white. The approach to the runway is illuminated by the approach landing system (ALS) comprising light bars and/or flashing strobes. The start of the runway is marked by green threshold lights, while the end of

GETTING THE HEIGHT RIGHT

The angle at which the airplane approaches the touch-down zone is critical, and to ensure that this is correct a system of lights called the Visual Approach Slope Indicator, or VASI, is used in many airports to provide a visual flight path. There are differing versions of this, but in principle it consists of two pairs of lights positioned at before and after the ideal touch-down point. Each of these lights projects a white light in the upper part of the beam and a red light in the lower part. They are angled so that a pilot on the correct trajectory sees the first set of lights as white and the second set as red. If all lights are red, then the plane is too low. If they are all white then the plane is too high.

The Precision Approach Path Indicator, or PAPI, is a refined version consisting of a row of four dual-beam lights positioned beside the runway. When the pilot is on the correct glide path the two lights on the left are white and the two on the right are red. Three or four red mean the plane is slightly low or very low respectively, while three or four white mean it is slightly high or very high.

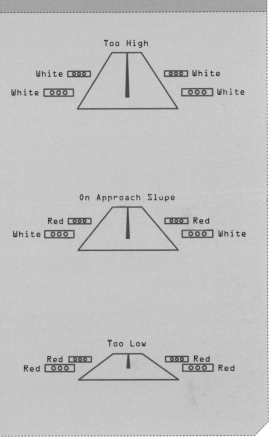

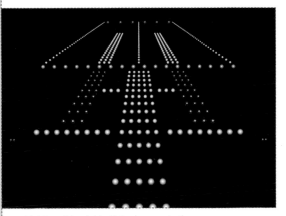

Lights guide night pilots down onto the runway.

the runway and any obstructions are indicated by red lights. The touch-down zone is highlighted by bars of white lights on either side of the centerline line, which is marked by white light, turning red at the far end of the runway. White or yellow lights run along the edges of the runway, and blue lights line taxiways.

AIR TRAFFIC CONTROL TOWER

WHAT?
A tall airport building with a control center at the top overlooking the aerodrome.

WHERE?
All international airports, and many large national airports, have an air traffic control tower.

DIMENSIONS
Many air traffic control towers are more than 330 ft (100 m) tall.

The most prominent building on the aerodrome at a major airport is the air traffic control tower. This is the nerve center from which all movements of aircraft (both in the air and on the ground) and airport vehicles are monitored and controlled. The air traffic controllers who work here have the responsibility of ensuring that all aircraft and their passengers arrive and depart safely.

Although many airports are "non-towered" and rely on standard procedures and radio communication for safe landing, taxiing, and takeoff, when the volume of air traffic using the facility reaches a certain level an air traffic control tower is essential. The control room atop the tall tower usually has windows all round it to give controllers a 360° view of the airport and all the planes and vehicles moving on it, but air traffic control (ATC) also has the most up-to-date electronic monitoring and communications equipment in order to manage three main areas of responsibility.

Clearance Delivery and Flight Data
Every commercial flight must submit a flight plan, detailing (among many other things) the flight number, time of departure, destination, route, and cruising altitude, and it is the task of Clearance Delivery to coordinate with other control centers along the route to make sure the plan can be carried out as requested. Before taxiing to the runway, the pilot must get approval for the flight plan from ATC Clearance Delivery, and sometimes the pilot will be informed of changes, especially if there is bad weather ahead or congestion in airspace or at airports.

The task of Flight Data is to make sure controllers and pilots are kept up-to-date with important information concerning such things as weather, delays, and runway closures.

Ground Control
A plane, a vehicle, or even a person in the wrong place at the wrong time can cause a disaster. Ground Control is responsible for all movements in all "air-side" areas except for the runways themselves, including the taxiways and holding areas, and some airports use surveillance radar to monitor ground traffic. Nothing can move without authorization, and it is Ground Control that gives pilots permission to taxi to the takeoff runway. Communication is maintained using radios or aviation light signals (see opposite).

Local Control

Also known as Tower Control, this aspect of air traffic control is concerned with the runways themselves, in close collaboration with Ground Control. Once a pilot is at the takeoff runway, it is Local Control that gives the clearance to take off, as well as instructing the pilot on the heading to take and the altitude to climb to. The Local Controller also clears aircraft for landing and, once they are down, directs them to an exit taxiway, at which point Ground Control takes over.

Keeping Track

When a flight plan is accepted, a piece of paper called a "flight tracking strip," which is like a flight plan in miniature, is printed out in the control tower. This strip is physically passed from Clearance Delivery to Ground Control to Local Control as the plane makes its way from the departure gate and into the air. At this point a digital version is passed to the next level of control, which passes it from one airspace sector to the next along the flight route.

At the same time, the plane is being tracked by radar. Each plane has a transponder, a radio transmitter that sends out a coded signal so that the moving image of the plane on the radar screen is accompanied by a "data block" that tells air traffic control the plane's call sign, altitude, and ground speed.

AVIATION LIGHT SIGNALS

If a pilot's radio fails or a vehicle has no radio, air traffic control can communicate from the tower using a light gun. These are the signals used.

Color and Type of Signal	Aircraft on the Ground	Aircraft in Flight	Vehicles, Equipment, and Personnel
Steady green	Cleared for takeoff	Cleared to land	Cleared to cross, proceed or go
Flashing green	Cleared for taxi	Return for landing, followed by steady green at appropriate time	Not applicable
Steady red	STOP	Give way to other aircraft and continue circling	STOP
Flashing red	Taxi clear of the runway in use	Airport unsafe, do not land	Clear the taxiway or runway
Flashing white	Return to starting point on the airport	Not applicable	Return to starting point on the airport
Alternating red and green	Exercise extreme caution	Exercise extreme caution	Exercise extreme caution

COMMUNICATION

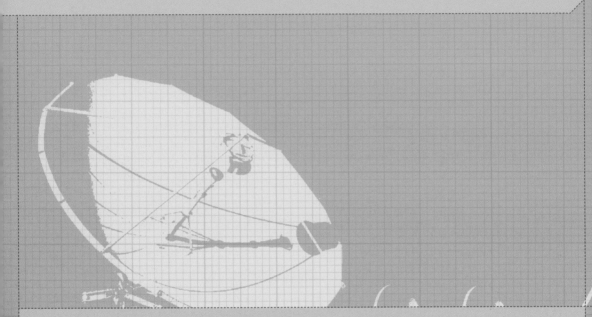

A BRIEF HISTORY OF COMMUNICATION

3000 BCE: The Egyptians use papyrus to record information

105 CE: Paper is invented in China by Tsai Lun

1450: The printing press is invented by Johannes Gutenburg in Germany

1680: The "penny post" is launched in London, costing one penny for parcels up to 1 lb (500 g)

1830 In the UK, mail is carried between Liverpool and Manchester by train

1838: American inventor Samuel Morse creates his eponymous code

1839: The first electric telegraph is laid in the UK

1851: A telegraph cable is laid across the English Channel

1866: A cable is successfully laid across the Atlantic

1876: The first telephone call is made by Alexander Graham Bell

In the urban environment, there is no escape from the constant flow of information. The endless communication means that we close our eyes and ears to much of it. This chapter draws your attention to the infrastructure that transmits the information.

People love to talk. The telephone has made this possible across large distances. The wires and cables that enable us to communicate criss-cross nations and travel under the seabed. The masts and towers supporting this network have become familiar features on the landscape. Satellites orbiting up above now carry a large part of the communication burden; this space-age technology brings entertainment and information into our homes.

The rise of the cell phone has made conversation possible at any time and in most places. This chapter looks at the masts that transmit digital signals between cell phones and the many disguises they adopt. The visual communication of roadside advertising is also examined.

1888: Heinrich Hertz discovers radio waves

1889: The 24-sheet "billboard" is used for the first time in France

1895: 21-year-old Guglielmo Marconi makes the first successful transmission using radio waves

1908: Henry Ford introduces America to the Model T car

1926: The first transatlantic telephone call is made

1936: The world's first public television broadcast is made

1946: The first commercial wireless telephone call is made

1951: Direct dial telephone is offered from New Jersey to 11 US cities; this service rapidly takes off

1962: Telstar, the first communications satellite to transmit telephone and television signals, is launched

1986: The earliest form of "Internet" emerges in the USA

BILLBOARDS

WHAT?
Large outdoor advertising signs that are provocatively designed to capture attention.

WHERE?
Located alongside busy roads and highways, or on the sides and rooftops of buildings.

DIMENSIONS
A standard "poster" or "30-sheet" size is 12 inches (3.6 m) high and 24 in (7.2 m) wide. A company in Kuwait created one nearly 3,200 ft (1,000 m) long to advertise the launch of a shopping mall.

Billboards have come a long way since the ancient Egyptians of Thebes first used a poster to advertise the offer of a reward for the capture of a slave. Their aim, however, remains the same—to stand out from the surroundings and capture the attention of passers-by. Older billboards from the first half of the 20th century can still be seen painted onto the sides of houses, although these have been pushed aside by the electronic advertising giants of the 21st century.

Birth of the Billboard

Since the first 24-sheet billboard was displayed in Paris in 1889, the advertising world has never looked back, finding bigger and glossier ways to get their products noticed.

The predecessor of the billboard was the poster. At first, the poster was intended simply to convey information to the public. However, the invention of more sophisticated printing methods in the late 18th century helped make the poster something of an art form and gave new depth to advertising. The next leap forward occurred in the early 1900s with the introduction of the electric billboard, lighting up dark cityscapes with a bright, colorful new mode of communication.

As the number of roads and highways multiplied, so did the number of billboards along their edges, which grew in size and were positioned with increasing prominence. The art of the billboard changed with the times, boosting the morale of depression-hit US citizens during the 1930s and distributing propaganda for the Red Army in China. They also played their part in the creation of contemporary celebrity culture, using famous media figures to endorse brands from the 1960s onward.

Billboards are now evolving from static posters to moving screens. Despite their expense compared to a printed sheet, their increased impact and versatility, and therefore potential revenue for the client, means that they are likely to become more commonplace.

ROAD SAFETY

Critics of billboards claim that they are a distraction to drivers and cause accidents. Insurance and traffic safety experts, however, have failed to find any conclusive proof that billboards are a factor in road traffic accidents, although the reluctance of drivers to admit responsibility for a crash calls into question the validity of the available data.

A 1994 billboard campaign in the UK for a brand of lingerie was credited with causing a number of car accidents among male drivers, although this is no more than an urban myth. The advertisements were toned down in the USA, but still succeeded in catching the eye.

In rural areas, the billboard sighted in the distance may help to relieve boredom. Along monotonous sections of highway some drivers have been known to fixate on the novelty slogan ahead and drive straight into it, a phenomenon known as "highway hypnosis."

Large billboards of the type seen alongside highways typically consist of a steel frame supported on steel poles. The frame supports the "facing," the backing material on which the artwork is fixed, and any associated lighting apparatus. Due to their large surface area and often exposed position, they must be able to withstand the worst that the weather can throw at them.

A street in Beijing, China, covered top-to-toe in billboard advertisements.

UTILITY LOCATION MARKINGS

WHAT?
Marks in colored spray paint, using letters, lines, and arrows.

WHERE?
Along the streets and sidewalks, wherever road maintenance or repairs are due to begin.

DIMENSIONS
Variable, as wide as a spray can, as long as a street. Their life span is brief, in New York at least 48 hours but no more than 10 days (excluding weekends and civic holidays).

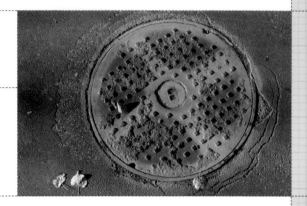

Utility location markings label above ground anything underground that is liable to get damaged in the course of carrying out roadworks. The gas pipes, water pipes, and electricity lines that run beneath our feet have, over time, become a complex network. Before a section of road or sidewalk is dug up, these utilities must be clearly marked so that disruption to the network is minimized. Accidental damage can be dangerous and extremely costly; for example, rupturing a water main could cause severe local flooding and contaminate the drinking water to thousands of people.

A stretch of water pipe exposed by roadworks.

For reasons of convenience, safety, or aesthetics, public utilities have been run underground for centuries. Gas companies in the UK successfully argued that if the wagons of their coal competitors could travel along the public right of way making deliveries, then they should be able to transport their product along the same streets. Roadwork signs went up in 1807 along Pall Mall in London, where the first gas pipe was buried under the street. Since then, streets have accommodated almost every new utility, creating a network of cables, pipes, and conduits.

Underneath our cities you will find major high-voltage electricity cables, smaller telephone and television wires, and pipes bringing in gas and water, and taking away waste and storm water. Each

AMERICAN UTILITY LOCATION MARKINGS: COLOR CODE

Utility marking color codes are not internationally regulated, although they are similar in countries across the world.

Red	Electric power lines, cables, conduit, and lighting cables	**Blue**	Drinking water
Orange	Telecommunication, alarm or signal lines, cables, or conduit		Reclaimed water, irrigation, and slurry lines
Yellow	Natural gas, oil, steam, petroleum, or other gaseous or flammable material	**Pink**	Temporary survey markings, unknown/unidentified facilities
Green	Sewers and drain lines	**White**	Proposed excavation limits or route

streetlight, display board, and traffic light has its own underground electricity supply. All of this must be accommodated alongside road and rail tunnels, mass transit systems, and national defense communication lines that make up the underground fabric of our lives.

Locating the Utilities

Digging up a road provides a timeline in pipe technology, since new pipes are often laid without removing the old ones. Gas mains have been made from cast iron, tin plate, steel, and now plastic. Different detection and location methods must be used to cover the variety of materials used by our public utilities. Electromagnetic equipment with a transmitter and a receiver is often used to find metal pipes and cables. For plastic or concrete pipes, radio location or ground-penetrating radar is used. Locating services by these means is necessary because maps often don't provide the pinpoint accuracy

required. This is a particular problem in older cities, since maps may be missing, incomplete, or out of date.

Before digging up the street, engineers mark the location of underground utilities via standardized system of color markings (see above).

In the USA, utility location is taken so seriously that its provision is obligatory by law. Contractors that are due to begin roadworks must notify the relevant authority in advance so that utility markings can be made. Failure to do this can result in a fine or even a criminal charge, particularly if such negligence causes a major incident. Given the dangers of accidental damage, a few utilities, mainly petroleum products, are permanently marked with short posts or bollards. The spray paint markings generally appear a few days before the diggers, and disappear under new paving once the work is complete.

THE MAIL

WHAT?
A network of people, bicycles, vans, trucks, trains, planes, and ships all dedicated to delivering a letter from A to B, no matter the distance between them.

WHERE?
In urban areas there's probably a mail box at the end of your street. The second largest civilian employer in the USA is the Postal Service, which also runs the world's largest civilian vehicle fleet.

DIMENSIONS
Truly global. The mailman may walk down your drive every morning or, if you are out in the wilderness, you might take delivery just once a month, but the network that allows both to happen is vast.

For the urban spotter out in the field, it is impossible not to find some evidence of the mail service. Most mail services around the world deliver to domestic addresses once a day, six days a week. To achieve this, almost every form of transport invented has been utilized and adapted in order to make the delivery of mail more efficient.

Something Old, Something New
The mail service is a mix of old and new. Although it is attempting to integrate new technology, it has also had to upgrade and adapt older infrastructure. From the ports and airports, a fleet of trucks will transport the mail to the sorting office, to be prepared for delivery the next day. In the past, all mail was sorted carefully by hand; now, automated machines have been introduced to some mail services that can read the type or handwriting on the front of each item of mail. Nevertheless, wire baskets and pigeon holes remain, used by the mail carrier to order the post for their route.

The logos and branding of postal services around the world are chosen with great care; the mail service needs to be trusted by its customers and branding plays a part in this. Understandably, mail theft is taken very seriously. In the USA, it is a violation of federal law for anyone other than the receiver to open mail.

A mailman doing the rounds in Curitiba, Brazil.

The Humble Letterbox

Post box, collection box, pillar box, drop box; they might be different colors and shapes but they are all physical boxes into which the public can drop outgoing mail to be collected by the country's postal service.

Boxes located outdoors are designed to keep your mail secure and protected from the elements. Post boxes are generally of a solid, robust nature, with the entrance slot designed so that post can be deposited safely, without the risk of theft. Collection is possible only with use of a key, which is held only by authorized personnel.

So durable are the UK designs that, after a1996 bomb attack in Manchester, one of the few things to survive unscathed was a pillar box from 1887. Post boxes are generally brightly colored for identification; the colors may signify different services, from regular to express. Color can also indicate different mail service providers.

This pillarbox in Manchester, UK, bore the brunt of a bomb attack in 1996.

INTERNATIONAL POST BOX COLORS

Red	Countries include Australia, Canada, Denmark, Hungary, Iceland, Italy (domestic post), South Korea, Japan, New Zealand, Romania, Spain (express mail), South Africa, UK
Yellow	Countries include Australia (express post), Brazil, Bulgaria, Finland, France, Germany (Deutsche Post), Slovakia, Slovenia, Spain (regular mail), Sweden
Blue	Belarus, Germany (many private postal companies Italy (air mail only), Portugal (first class only), Sweden (local mail), Russia, USA
Green	China, Hong Kong (red before 1997), Taiwan, Ireland, some heritage boxes in the UK, notably Stoke-on-Trent, Rochester, and Scunthorpe
Orange	Czech Republic, Estonia, Indonesia, Netherlands (TNT boxes; red before 2006)
White	San Marino
Gray	Philippines

TELEPHONE SYSTEM

WHAT?
Originally a conversation via a wire, now a complex infrastructure of handsets, wires, poles, masts, and exchanges.

WHERE?
Worldwide, there are over 1.26 billion telephones connected by fixed lines. Using metal cables, radio links, fiber-optic cables, and satellites, it is possible to pick up a telephone in your home and call someone on the other side of the planet.

DIMENSIONS
Although the associated telecommunications infrastructure is vast and covers large swathes of the planet, it is not always evident to the uninitiated.

Landline telephones are among the simplest machines in your home and, until the advent of cell phones, they remained relatively unchanged since their invention over 125 years ago. The first telephone call between the east and west coast of the USA took place in 1915, and in 1969 President Nixon was able to call astronauts orbiting the Moon from the comfort of the Oval Office in Washington.

The telephone has been an important part of our lives for several generations. The sheer scale of the telecommunications network means that, for the urban spotter, telephone equipment is always in sight. Look out for past and present company name plates on telegraph poles, manholes, exchange buildings, and public telephone boxes. Telegraph or utility poles remain much the same although their lines have been upgraded from plain copper to fiber-optic.

Local exchanges have evolved from a building with a switchboard operator individually transferring all the calls in a small area to a small gray metal box on the side of the road filled with miniaturized switching equipment. The main exchanges receive calls from a number of small local exchanges and transmit them via radio links or fiber-optic cables to microwave towers and international exchanges.

When you talk on the telephone, the microphone turns the sound waves of your voice into a variable electrical current. This electrical current is converted by the receiver back into sound at the other end of the line.

For 125 years, the telephone system has delivered information and news and kept people in touch, even when they are far apart. The rise of cell phones and computer video conferencing will eventually make the hard-wired handset in our home become a museum piece.

THE GLOBAL TELEPHONE NETWORK

The electrical current into which your voice is converted when you speak into a telephone has to go through quite a journey before it is turned back into sound at the other end.

First it travels from your telephone handset through the wall socket and out to the nearest telegraph pole. From there, it travels from pole to pole along wires to the local exchange, and then on to the main exchange, where it is transmitted to a relay tower via a microwave radio link. The relay tower bounces it to an international exchange then up to space to a communications satellite. If your friend is on a cell phone, it comes back down to earth via the international exchange to a mobile exchange. The final step of its journey is to the mobile handset via a base station radio link. Each of these stages is visible to an urban engineering spotter—even the satellite, if you're lucky.

Telephone poles stretch out across a field in rural Slovakia.

CELL PHONE TOWER

WHAT?

An elevated location where electronic communication equipment and antennae are mounted as part of a cell phone network. In order to pick up a signal on your cell phone, antennae are fixed to the top of a tower. Although they don't rely on line-of-sight technology, buildings and geography will limit their effectiveness.

WHERE?

In suburban areas. Sites may be spaced 1–2 miles (1.6–3.2 km) apart. In order to handle the number of calls, they may be as close as ¼–½ miles (0.4–0.8 km) apart in crowded inner cities.

DIMENSIONS

Many telecommunication companies will share locations, each providing their own equipment but mounting their individual antennae on the same mast. Like the cell phones themselves, the size of the equipment used has shrunk.

Cell phones allow communication between any two points on the planet, provided there is network coverage. The capacity of a cell phone site limits the number of calls that can be made at any one time; therefore, the density of cell phone masts in a given area will reflect the number of potential users it contains. A base station enables cell phones to switch from one mast to another, mid-call, depending on the strength of signal, available mast capacity, and location.

Everyday Landmarks

A cell phone tower is typically a steel pole or lattice structure that rises hundreds of feet into the air. The antennae at the top, which are the most visible elements, are linked by a thick set of cables to radio transmitters and receivers at the base. These cables also carry the mast's power supply. Masts will be heavily grounded to avoid damage from lightning strikes.

As cell phone usage took off and the number of masts increased, so did the number of complaints. The companies responded by attempting to disguise new cell phone towers as everyday and familiar parts of the natural environment. The results have often been highly imaginative.

Combining cell phone technology with existing landmarks reduces their visibility and subsequently the number of complaints they generate. Antennae can be found mounted on chimneys, grain silos, bell towers, and even crosses outside churches. They have been disguised with fake brickwork or a coat of paint, and owners of properties that house antennae receive money from the telecom company.

Privacy Issues

The constant communication between the cell phone

and the tower allows individuals to be tracked on a scale never before imagined. The handset, whilst in your pocket or bag, emits a "roaming signal" to locate the nearest tower. The surrounding cell towers pick up the roaming signal and calculate which tower is best able to provide coverage. As you travel and the phone changes location, the tower continues to monitor the strength of the signal, switching the phone to another tower when the signal becomes weak. The location of you and your phone can easily be determined by comparing the signal strength received by nearby towers.

What the service provider does with this information is part of the ethical debate that any advance in technology brings. The right of politicians, law enforcement agents, and telemarketers to access this data has to be balanced against the privacy rights of the individual. Cell phone towers offer the freedom to make calls whilst moving around, but the cost is that somebody, somewhere can always find out where you are.

CELL PHONE "TREES"

The transformation of the oblong, gray cell phone antenna into a palm tree is brave, imaginative, but only rarely successful.

Telephone companies have a range of mast "species" to match the variety of local flora. A "monopalm" is a mast disguised as a palm tree, and the rarer *"Pseudpinus telephoneyensis"* is a mast masquerading as a pine tree. The alternative is simply to paint the whole lot green and hide it in the middle of a forest.

The distinctive silhouette of a cell phone tower among the rooftops of Casale Monferrato, Italy.

TELEPHONE BOOTH

WHAT?
A home for a telephone for use by the public or the police.

WHERE?
They are used worldwide.

DIMENSIONS
The iconic British red telephone box is 3 ft by 3 ft by 8 ft 4 in (1 x 1 x 2.5 m) in size and weighs 1,600 lb (750 kg). In other countries, the telephone is positioned beneath a canopy, which is fixed to the top of a pole.

The telephone booth is a quiet public oasis in a busy urban world. Whether it's a box in the UK or a booth in the US, Canada, or Australia, this small structure houses a telephone available for anyone with the means to pay. While they are red in the UK, yellow or blue in France, and green in Cyprus, they can appear in just about any color.

Early telephones suffered from sound decay over long distances so "silence cabinets" were constructed to help users hear clearly, and also to make themselves heard without disturbing the neighbors. An outdoor booth usually has a door to maintain privacy and protect the equipment and the caller from the

PAY PHONES: PAST, PRESENT, AND FUTURE

1889	The first public coin-operated telephone was installed in a bank in Hartland, USA. Coins were deposited after the call was made.
1898	The Western Electric No. 5 Coin Collector was introduced, which was the first to insist on payment before placing the call. This became standard practice until 1966.
1902	The USA had over 81,000 pay telephones.
1905	A Cincinnati street saw Edison's first outdoor paybox installed. People were initially reluctant to make private calls on a public street.
1950s	Glass telephone booths start to replace wooden ones in the USA.
1957	The drive up and call from your car pay phone appears in Chicago.
1960	Telephone company Bell Systems installs its millionth pay phone in the USA. The number peaked in 1996 at 2.2 million and has declined thanks to the rise of the cell phone.
1980s	The rise of the phone card—no coins necessary, just a prepaid plastic card.
2006	New York phone booths are adapted to offer a wireless Internet connection.

POLICE BOXES

The first police telephone was installed in Albany, New York, in 1877, just one year after Alexander Graham Bell invented the device. Other cities were quick to follow. For the exclusive use of the police, a key enables access to a small box containing a direct line telephone. A more sophisticated arrangement allowed the public to signal up to eleven different alarms via a dial mechanism including "Police Wagon Required," "Thieves," "Forgers," "Murder," and "Drunkard."

The large wooden police boxes still occasionally seen in Britain's cities were introduced in 1923. Containing a table, stool, first aid kit, emergency supplies, and the all-important telephone, they also featured a blue light on top that would flash to indicate that any policeman in the vicinity should contact the station.

Offering a similar service to police boxes, "help points" can now be found in cities that allow the public to contact the police or transport authorities. The call is often patched through to a central control room, where the caller can be observed via CCTV. From this, the control room operator will be able to judge the appropriate response.

weather. The clear windows allowed the public to see whether the booth was in use and discouraged vandalism.

Pay As You Talk

Modern pay phones give the user many payment options: coins, pre-paid cards, credit or debit cards, or even calling the operator to reverse the charges, although this last one is prohibitively expensive. The public location of the phone booth exposes it to the elements and also to unwanted attention from people; acts of vandalism have necessitated simple and robust designs that can be repaired easily, and relatively inexpensively. Phone booths across the world still carry the symbols of long forgotten companies, the hardware having outlived the original owner.

From the 1970s booths became less common in the United States; in their place came non-enclosed pay phones, easier for disabled access and discouraging lengthy calls in high-demand areas such as stations and airports.

Alternative Uses

Although still fairly common, the rise of the cell phone since the late 1990s has seen a decline in the number of phone booths needed in busy locations. Many call boxes in rural areas have been retained for emergency use only. However, cell phone use is unlikely to force the complete disappearance of the phone booth. The prime location of these public telephones in busy places, close to roads with power supplies, offers a flexibility crying out for imaginative reuse. Trials have seen booths and boxes converted to charging points for electric cars in Madrid, providing wireless internet access in Manhattan, and used as billboards in London.

MICROWAVE TOWER

WHAT?
A tall, thin structure covered in dishes and aerials that send out and receive radio signals across vast distances.

WHERE?
Up on high ground. As microwave towers work by relaying information directly to one another, you'll find them hundreds of miles apart but always in sight of each other.

DIMENSIONS
Microwave towers are very tall. With no obstructions around, they are a distinctive landmark covered in pointy antennae and round satellite dishes up to 13 ft (4 m) across. These are all carefully aligned with other towers on the horizon.

Microwave towers transmit both digital and analog signals over long distances, sending telephone calls, radio signals, and relaying television programs to transmitters. Cutting-edge technology from the 1950s, microwave links involve line-of-sight communication technology. Affected by the environment, they are vulnerable to rain and sensitive to solar proton events and even high pollen counts, which affect the strength of the signal received. With the explosion of the Internet, the bandwidth available using microwaves became insufficient. Despite this, and the fact that fiber-optic cable and satellite dishes took over some of their functions, the towers still remain. With antennae upgraded and extra dishes added, they are now used for cell phone transmission.

The Rise of the Towers
The golden age of microwave tower construction occurred in the 1960s, as the threat of an escalation in the Cold War hung over Europe and the USA and civil engineers and architects pushed the boundaries of material technology to produce ever more advanced designs. The legacy of this is clear to see; in many capital cities around the world, the communications tower is the tallest, most notable structure. Reinforced concrete, steel lattice frames, and large expanses of glass are a common sight,

along with high-speed elevators. Perhaps less so are the revolving restaurants with dramatic views that can be found at the top of some towers.

There are many steel lattice towers supporting communications equipment. It is a simple but effective design that combines elegance with great strength, low weight, and wind resistance. They are also economically built, using less material than other designs. Of the lattice towers, those made up of triangular cross-section are most common.

Bombproof

Towers built during the Cold War were designed to withstand nuclear attack and the subsequent fallout, with concrete bunkers constructed close by to protect personnel and equipment. The microwave "horns" were covered with a protective shield to keep out not only the elements but also radioactive fallout. The buildings were shielded with copper to protect the equipment against the electromagnetic pulse associated with a nuclear explosion. Concrete walls 1 ft (30 cm) thick protected the vital electronics and people inside the base installations of these towers. Thick copper grounds went deep into the bedrock beneath each tower. Fallout showers, backup generators, and sleeping facilities all existed to enable the network to function in times of war.

The Communications Network

The central part of a communications network is referred to as its backbone, which joins major centers via high-capacity communication links. The Internet operates along similar lines: it also employs microwave transmission towers positioned in a line-of-sight network to move large amounts of data over long distances. Given that large objects such as mountains can potentially disrupt this network by blocking the microwave signal, the positioning of each tower is carefully chosen. When done well, microwave transmission can be remarkably efficient and effective; in 1956, it took only 14 towers to transmit a signal the length of the UK. For the keen spotter, microwave towers often support a wide variety of receiving and transmitting equipment, from the early parabolic reflectors and the characteristic 1960s horn antennae to modern satellite dishes and cell phone transmitters. If a tower is covered with an ugly variety of ill-fitting hardware, it is probably because the original antennae have been retired and replaced with newer equipment that the tower can support but was not designed for.

THE "SECRET" TOWER

The Post Office Tower in London, UK, has been a landmark since construction began in 1961. The 620-ft (189-m)-tall cylindrical tower is made from 14,300 tons (13,000 tonnes) of concrete, steel, and glass. Initially, the first sixteen floors were for technical equipment and power, followed by a 114-ft (35-m) section for the microwave aerials. Above that were six floors of suites, kitchens, technical equipment, and finally a cantilevered steel lattice tower. Its shape was designed to reduce wind resistance and give it stability and style. The revolving restaurant at the top offers a complete aerial view of London every 22 minutes.

The tower has become a key component in Britain's new microwave telecommunications network. Despite the availability of postcards of this central London landmark, complete with its own restaurant, the tower was unmarked on official maps until 1995.

SATELLITE DISH

WHAT?

If you are hoping to receive transmissions from deep-space alien lifeforms, or would simply like to watch television, a dish-shaped parabolic antenna will pick up the necessary electromagnetic waves.

WHERE?

Domestic dishes can be found perched on rooftops or bolted to walls. Large radio telescopes and satellite farms are an unmissable landmark in remote areas.

DIMENSIONS

A small dish might be no more than 12 in (30 cm) across. In Puerto Rico, the Arecibo radio telescope is 1,001 ft (305 m) in diameter and built into the ground.

A curved metal dish picks up incoming radio waves and reflects them to a focus point, the antenna, which is positioned above the center of the dish. From there, the information is delivered to the television or satellite box via a wire. Larger radio telescopes work on the same principle as the domestic dishes, but on a far bigger scale.

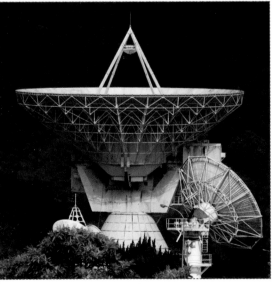

A satellite dish in northern Germany.

Terrestrial or Otherwise

Many objects in the universe emit radio waves, some of which are extremely weak. When a large dish collects these waves and reflects them onto the antenna, they are converted into an electronic signal. This signal is sent to a computer, where it is combined with signals from other dishes to give us an idea of what is out in space.

Television signals are far stronger than those detected in space. This means that domestic dishes can be much smaller and are therefore more affordable. Satellite television gives individuals a wide choice of channels from different sources—many of these free-to-air—and often picks up broadcasts from different countries. This freedom of information is particularly prized by those living in countries where the media is strictly controlled.

In the US, cable TV stations relay satellite TV to subscribers from their own central satellite dish receiver via a coaxial cable. In Europe and Japan, a more direct method is favored, whereby signals are sent directly to a viewer's dish at their home.

The basic satellite dish consists of a parabolic reflector made of fiberglass or metal, with a protruding steel feed horn and amplifier in its middle. A domestic dish will usually be fixed to whichever side of the house provides the clearest view to the transmitting satellite. Once you have spotted one, you will see a whole line of them along the street, all facing in the same direction.

Dish Positioning

In order for a dish to pick up a signal it needs to be aligned with the satellite up in space. The alignment settings vary according to the geographical location of your dish and can be worked out using a complex series of calculations that takes into account your latitude and longitude relative to the satellite. Alternatively, you can check online. There are three things to consider when positioning a satellite:

Azimuth: This points the dish in the right direction. In the UK, this varies between 142 degrees in Northern Ireland and 150 degrees in the most easterly part of England. There are satellites every three degrees, so it's getting pretty crowded with satellite choices.

Elevation: How far up to angle the dish. Most dishes are offset so that the Low Noise Blocker, which receives and channels the signal, doesn't point directly at the satellite.

Skew: This involves rotating the Low Noise Blocker to fine-tune the signal.

BIG UGLY DISHES

Television receive-only (TVRO), or "Big Ugly Dish," is a North American term for an almost redundant form of television reception, the C-band satellite. Sold from the late 1970s, these large domestic dishes were at the cutting edge of technology, bringing an increasing number of television channels to areas beyond the broadcast range of small local TV stations. At their peak, large numbers of free-to-view programs were available, although at 6–12 ft (1.8–3.6 m) in diameter and made of fiberglass, they were often considered an eyesore. Gradually subscription encryption was introduced and broadcast satellites improved, leading to the availability of smaller and cheaper domestic dishes.

For the owner of the Big Ugly Dish the flip side of greater television choice was an increase in neighborhood disputes, with calls made for legislation and zoning permits. Many dishes, although long since disconnected, still occupy their original sites. Too large to remove easily, they are a monument to redundant technology.

WASTE

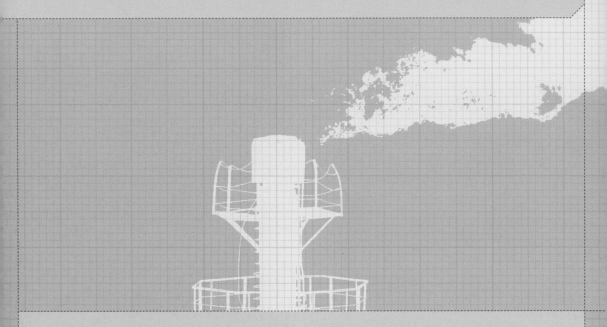

A BRIEF HISTORY OF WASTE

3000 BCE: The earliest recorded landfill sites are in use in ancient Greece

1185: The earliest known example of a chimney in the UK is built

1348: Street cleaning laws are introduced in Florence, Italy

1554: In France, 800 carts remove trash to the countryside twice a day

1835: The Adams soap factory in Birmingham, UK, builds the world's tallest chimney at 312 ft (97.5 m)

1884: Only two towns in England collect refuse at public expense

1858: The stench of sewage from the River Thames in London is so bad that the UK parliament has to be suspended

1865: In Gibraltar, Michigan, the first public waste incinerator is built to destroy all refuse

1875: The UK outlaws the use of children to clean chimneys

In this chapter we cover the good, the bad, and the smelly ways to deal with the waste products of our lives. In the wake of human progress, a great deal of waste often remains; it's something most of us would rather not deal with. This chapter looks at the specialized engineering solutions that, thankfully, do the bulk of the work.

Manholes are often the only clue to the subterranean system of sewers and storm drains beneath the roads and sidewalks. This chapter examines the places that most people see rarely, if at all: the sewage treatment plant; combined sewers; and the incinerator, to name but a few.

The amount of waste that we generate in our day-to-day lives presents a significant logistical and environmental problem. The waste transfer station and landfill site have been designed to address this.

Throughout history people have made money from other people's waste. This century's greatest engineering challenge is finding new solutions in handling our unwanted products that make economic and environmental sense.

1890: The first sewage treatment plant to use chemical precipitation is built in Worcester, Massachusetts

1913: The Activated Sludge Process currently used in sewerage plants is invented

1923: Construction begins at Jones Island, Lake Michigan, on the first large-scale activated sludge plant

1956: The UK introduces the Clean Air Act, which requires businesses to use tall chimneys

1973: Plastic bottles start to be recycled. Due to rising energy costs recycling rates rise across all areas

1987: The world's tallest chimney at 1378 ft (420 m) is built in Kazakhstan

2005: Computer cellphone recycling schemes are well established. Battery recycling helps bring down the cost of nickel and cadmium

2010: The global price of recycled paper rises from $120 to $217 in six months

CHIMNEYS

WHAT?

A structure for releasing waste gases into the atmosphere. If it is attached to a ship or a locomotive, it does the same job but is called a funnel or smokestack.

WHERE?

Whether part of a building or an independent structure, by interrupting the skyline they draw attention to themselves and often become prominent landmarks.

DIMENSIONS

Size varies according to usage. The world's tallest chimney, at 1,400 ft (420 m), can be found at the GRES-2 power station in Kazakhstan.

Many of our industrial processes involve burning, cooling, or having to discharge gases high above our heads so that we don't breathe their waste. Add to this private fireplaces in homes around the world and it's no wonder you'll be able to spot chimneys all over the urban landscape.

Humans have understood for a long time that the fumes coming off fires aren't pleasant to breathe. The simple solution, as we've found, is to force them up and away through a pipe made out of a material that can withstand the heat. During the Industrial Revolution of the 18th and 19th centuries, larger chimneys began to proliferate, and many of these can still be seen across Europe. However, you won't spot any brick chimneys in areas that experience earthquakes as they tend to crumble when shaken.

Safe Dispersal

Chimneys make use of the natural "draft" effect. Air outside of the chimney pushes the less dense hot air up the flue. The taller the chimney, the greater the draft. The height of the chimney is also key to its ability to disperse the waste gases. Therefore, as health concerns increased, so did the height of industrial chimneys. Reinforced concrete replaced brick as greater structural strength was needed for these tall chimneys. The concrete tube that you see attached to a modern boiler or furnace may just be the protective shell of the chimney, containing a number of flue liners inside. You can even spot large industrial chimneys at night. Above a certain height a chimney must display a series of warning lights toward the top to prevent aircraft from flying into them.

Unlike industrial chimneys, most of a domestic chimney will be hidden inside the walls of the house, with only the chimney pot visible on the roof. The pot is added to extend the height of the chimney to improve its draft. Large houses with many floors and fireplaces will have many pots, often with decorative terracotta moldings or fancy twisting brickwork. The chimney can be finished off with a cowl, which keeps out the rain, adventurous squirrels, and broody birds. Similar wire caps were often used on steam engine funnels to capture sparks and prevent fires, particularly when the engine was driving a threshing machine or carrying a flammable load.

CHIMNEY PLUMES

Once you have spotted the chimney, have a look at what's coming out of it. The way in which fumes from a chimney are dispersed into the air depends on the weather conditions, particularly how cold and windy it is. Early in the morning, on a calm day, the smoke or steam leaves the chimney and is blown horizontally by the wind, slowly fanning out. If the sun heats up the ground quicker than the air above the chimney, the smoke can mix downward, which may lead to "fumigation" of the surrounding land. The most striking pattern is "looping," later in the day, when the air has warmed up and wind has strengthened, the plume can be pushed up and down in a wavy pattern, causing the smoke to form a loop.

Ornate chimney pots in Albufeira, Portugal.

MANHOLE

WHAT?

The manhole marks an entry point to our water, power, sewer, gas, telephone, and steam systems. The heavy metal cover hints at an intricate unseen network below.

WHERE?

In the street, beneath your feet. Wherever pipes, cables, and tunnels run underground, access points are needed.

DIMENSIONS

A 27-in (70-cm) round cover will weigh approximately 300 lb (130 kg). The shaft beneath will be approximately 40 sq in (100 cm^2).

We have buried pipes beneath our roads for thousands of years, from ancient Roman sewer systems to fiber-optic communications cables, all of them safely hidden out of the way. However, we need access to this underground network in order to maintain pipes and tunnels, connect cables, and trace breaks and faults. This access is provided by manholes.

A manhole is the opening to a man-sized shaft with a ladder or rungs leading from the surface to the inspection chamber below. It is covered with a metal or precast concrete plug to prevent accidental or unauthorized access. These covers have to be strong to allow vehicles to drive over them, and heavy, to prevent them being dislodged. In order to lift the cover, the authorized worker (or unauthorized urban explorer) inserts two keyed rods into the "pick holes" on the surface and lifts it up. Occasionally, covers are hinged.

Intelligent Design

The characteristic raised patterns on the covers were originally specified to prevent wagons, and the horses pulling them, from slipping. Today's motorized transport has less difficulty and the patterns have become more aesthetic. This has been taken to its extreme by the Japanese. Almost all municipalities in Japan have their own custom-made manhole covers, sporting a wide range of colorful images. Designs include images that reflect a region's culture, flora and fauna, and landmarks.

Prized Assets

If you see a manhole with no cover, it has probably been stolen. The large volume of metal used to make a heavy manhole cover means they are valuable to scrap dealers. The large-scale theft of manhole covers has been reported all over the world, often with many stolen in the same night. In Calcutta, India, more than 10,000 manhole covers were taken in two months. These were replaced with concrete covers, which were promptly stolen as they contained a metal reinforcement bar.

Don't confuse manholes with coal holes. The coal hole was introduced in the 19th century, when coal was widely used for heating, and can be seen set into the sidewalk in front of town houses.

A commemorative manhole cover in Albuquerque.

The grooved surface of a French cover.

A Japanese manhole cover depicting two firefighters.

A utility cover in Fredericia, Denmark.

A beautifully detailed example in Bergen, Norway.

An example from Windsor Castle, Windsor, UK.

WHY ARE MOST MANHOLES ROUND?

Reasons for the shape include:

- A round cover cannot fall through its circular hole, whereas a square one, if placed diagonally, could.
- The circular cover can simply be dropped on the frame without having to line up the heavy metal casting.
- Round castings are much easier to machine flat with a lathe.
- Round tubes are the strongest and most material-efficient shape for burrowing through the ground.
- A heavy round manhole can be more easily moved by rolling.

COMBINED SEWERS

WHAT?
A large pipe system that collects unsavory sanitary waste and the water runoff from the streets, mixes it all together, and discharges it far away.

WHERE?
Difficult to spot as they are almost entirely underground. They are generally gravity-powered so run downhill toward the treatment works or, during high flows, discharged directly into a body of water.

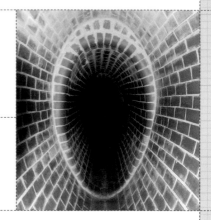

DIMENSIONS
Modern pipes are large, round, and unremarkable, but the older brick-lined sewers have dramatic, cavernous spaces with vaulted ceilings, staircases, and cascades of foaming brown water.

As our cities become more affluent, so the treatment of our effluent becomes more sophisticated. The combined sewer system carrying both sanitary waste and surface runoff is the original solution; having served us well for nearly 150 years, it is gradually being upgraded. This will ensure that no untreated waste will enter the environment.

The Big Stink
London in the 1850s was an extremely unpleasant and unhealthy place to be. Open sewers, discharging directly into the Thames, contaminated water supplies, and swarms of flies feeding on horse dung and excrement caused disease epidemics to sweep the rapidly expanding city. In the summer of 1858, a heat wave caused the smell to become so bad that the British parliament (which is next to the River Thames) was forced to act. The result was a network of combined sewers across the city that is still in use today. Within six years, the "big stink" of London was consigned to history.

Today, the sewerage system works in broadly the same way: domestic and industrial waste enters the pipes and, whether by gravity or pump, ends up at a treatment works, where it is treated before being discharged to the environment.

During severe wet weather, the rainwater runoff increases the flow rapidly. Wet weather flows can be between three and five times higher than in dry conditions, occasionally flooding the pipe network and exceeding the capacity of the treatment works, and increasing the risk of sewage backing up in homes and streets. The traditional, low-cost solution was not to build bigger treatment plants that would stand idle for long periods of time but, given that the sewage would be diluted by the large quantity of storm water, allow direct output without treatment to a river or the sea.

Increased awareness of the environmental impacts of this method have led to combined sewers being upgraded or replaced to avoid waste being discharged without treatment during periods of high rainfall. Sewer effluent and storm water can be routed into separate pipes, although this is

WASTE

expensive and its installation causes major disruption. Alternatively, large holding tanks or tunnels can be built to store the excess waste until the treatment plant has the capacity to treat it. Methods to reduce both sewage and storm water runoff are also being encouraged. If rainwater can drain away naturally through permeable surfaces it doesn't need to enter the sewage system. Rainwater and "gray" water from baths and showers can be harvested and used for watering the garden rather than flushed away.

THE COMBINED SEWER

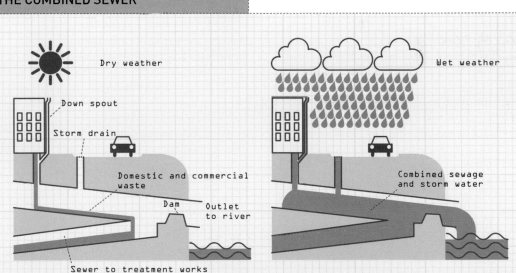

Water and sewage is only diverted to the outfall pipe during especially high rainfall, when the treatment works cannot handle the increased volume of water. The outfall pipe may discharge to a holding tank rather than discharge direct to a river or other body of water.

SEWAGE TREATMENT PLANT

WHAT?
They treat the output from your household drains to produce liquids and solids that won't harm the environment.

WHERE?
Follow the drains and your nose. As a large, unpopular neighbor, the treatment works will generally be found on the outskirts of town, hopefully downwind from your home.

DIMENSIONS
A series of large sprawling tanks and channels, often a little "sci-fi" in appearance, and with a distinctive smell.

Many processes in the sewage treatment plant are designed to mimic the process of waste being broken down by nature. In the engineering world, the filters, bacteria, and settling processes are given exciting technical names and carried out on a large scale. The end result is that the water, or effluent to engineers, can be safely released back into the environment and the solid waste, or sludge, can be disposed of or recycled as farm fertilizer.

Sewage treatment has evolved into a three-stage process:

Primary Treatment
The sewage flows through large tanks, where the heavy solids settle to the bottom and the oil, grease, and light solids float to the top. The solids and floaters are removed and the remaining liquid is subjected to secondary treatment. Mechanical surface skimmers, the most visually striking part of the system, collect the grease and fat from the top while the sludge from the bottom of the tank is pumped on for further treatment.

Secondary Treatment
The second stage removes all the dissolved and small suspended biological particles from the sewage. This is mainly done by bacteria, which require a blast of oxygen to help them eat through the rich nutrient soup and bind the less soluble parts together into "floc."

Tertiary Treatment
Known as effluent polishing, this improves the standard of the final discharge by filtering and removing the nutrients, nitrogen, and phosphorous that in large doses would damage the environment.

It is the early stages of sewage treatment that tend to release the smelly gases, with hydrogen sulfide being the major offender. Carbon reactors, a contact filter with bio-slimes or small doses of chlorine can be used by treatment plants to minimize the obnoxious gases reaching your nasal passages.

WASTE

Nineteenth-century British sanitary engineers led the way in the safe disposal of waste. One hundred and fifty years later, however, and large areas of the world are still without basic plumbing. Recognized as one of the most important factors in improving health in the developing world, it comes as a shock to realize that, while cell phones are widely used, effective sewage treatment is still not universal. In many developing countries, domestic and industrial waste water is discharged with little or no treatment back into the local environment. Parts of sub-Saharan Africa and South America have no strategy in place for dealing with the waste produced by a growing population. Above ground, Tehran in Iran may have gleaming office blocks and a 175-mile (280-km) highway network, but major parts of the sewerage system have yet to be built.

The World Health Organization estimates that waterborne diseases such as cholera and typhoid kill 25,000 people every day. The most effective preventative measure for these entirely avoidable deaths is not medical but structural: the installation of sewage systems and treatment plants.

AN UNLIKELY VEGETABLE PATCH

Tomato seeds cannot be digested by the human body and so find their way through the sewerage system to the treatment works. The dried sewage, of which they are a part, acts as a perfect fertilizer, enabling them to grow strong and healthy. Every year, sewage treatment plants across the world have a fine crop of tomato plants. However, they are not available in local supermarkets; along with the sludge they are removed from site.

With improvements to the treatment processes, it is now possible to reuse sewage effluent for drinking water, although Singapore is the only country to implement such technology on a production scale.

THE TREATMENT PROCESS

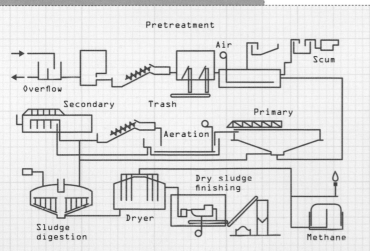

The sewage treatment plant receives private and commercial waste. The three stages separate solids from liquids, and filter and treat the different parts, before being released into the environment.

STORM DRAINS

WHAT?
Subterranean tunnels or simply large open channels that direct water from the streets at times of heavy rainfall.

WHERE?
Often underground, although trenches are sometimes created at the surface.

DIMENSIONS
The system is extensive and covers entire cities, stretching for many miles.

Our urban environment is one of flat, hard surfaces through which water cannot drain away. Without careful attention to detail, every passing rain shower would leave puddles of standing water on our parking lots, sidewalks, and streets.

Natural Inclination
In dry weather, the storm drain system remains unnoticed, but rainfall brings it to life. Gravity draws water to the lowest point on the street's surface, and civil engineers exploit this by designing streets with a camber that sends the water into drains at the edge of the roadway.

A storm drain in the gutter of a road in Illinois.

The problems posed by heavy rainfall are not new, and ancient civilizations employed familiar methods of channeling the water. Evidence of storm-water runoff systems such as drains and channels have been identified in the ancient Cretan cities of Minos and Knossos.

A storm drain system consists of a series of inlets, a piping network, and an outlet. A successful system will aim to collect a huge volume of water and transport and store it safely where it can do no harm. To prevent objects, such as debris and trash, getting into the system, gratings often cover the opening. Warning signs are sometimes used to warn people of their presence.

There are a couple of different inlet types. Side inlets are part of the curb—a vertical opening under the paving that captures the water flowing along the street gully. Grated inlets are set underfoot in the paving and have grids to prevent those large objects entering the sewer system. The spacing within the grid is wide enough to allow a sufficient amount of water to flow through it, while still protecting people from serious injury. Whereas the Romans used stone grates, the modern world prefers metal castings. A small lip is often added to collect objects that have entered the drain—particularly important if the object is valuable. Gulley pots in the UK act as a water-filled trap to prevent gases and rodents escaping via inlets.

The pipe system can be a simple network of round tubes; others are far more complex, incorporating waterfalls, staircases, balconies, and gross pollutant traps (GPTs—pits for catching rubbish) to labyrinthine effect.

Polluted Waterways

The traditional storm drain system (see the diagram below) poses environmental problems. While urban rainwater washes the streets clean of oil, dust, litter, and other waste, this can often be dumped, untreated, into lakes, rivers, or the sea. An alternative way of reducing the impact of heavy rainfall in urban landscapes is to leave more of the ground uncovered, thereby allowing the rain to infiltrate the soil rather than discharging through the drains.

"DRAINING"

Due to their large size and unfamiliarity, storm drains have been the focus of a subculture of urban explorers who go "draining." This involves the illegal exploration of storm drains beneath a city and is especially popular in Australia. Marginally less dangerous and far more pleasant than sewer exploration, draining is still a perilous pastime. Sudden heavy rainfall can flood a storm drain system with very little warning, causing explorers to be trapped underwater and drowned or killed by the impact of being flushed down a pipe at high speed.

STORM DRAINS

Unlike combined sewers, storm drains are separate from sewers. Catch basins take water from the streets to the drain, which has a large bore to handle sudden increases in water volume caused by heavy rainfall. They can be square, rectangular, egg-shaped, or round.

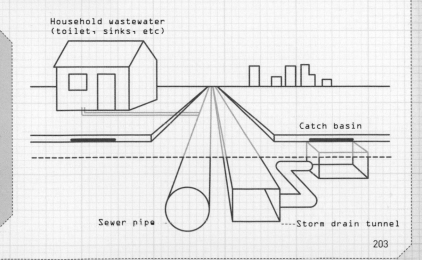

Household wastewater
(toilet, sinks, etc)

Catch basin

Sewer pipe

Storm drain tunnel

GARBAGE TRANSFER STATION

WHAT?
A site for the temporary storage, sorting, and transferral of waste for final disposal.

WHERE?
Never in the best part of town; just follow the garbage trucks.

DIMENSIONS
A single-storey, portal-frame building with a large concrete yard enclosed by a high boarded fence. Garbage transfer stations are distinguished by their complete lack of architectural merit; their design is purely functional.

Garbage transfer stations are temporary resting places in an unwanted object's journey from the domestic garbage can to its final resting place. Highly specialized local waste collection vehicles deposit their waste cargo prior to sorting and loading into larger vehicles. These large trucks will transfer the waste to the end point of disposal—incineration, recycling, reuse, or landfill.

Collection and Disposal
Domestic waste transfer stations are the collection points for residential waste gathered by the garbage trucks that roam our streets early in the morning. These vehicles have to be small enough to pass down narrow residential streets and, despite their compactors, can only carry a limited amount of waste. In some cases, waste from the trucks will simply be transferred to larger vehicles and transported direct to landfill or incineration. More modern stations don't just transfer waste; they also incorporate waste-handling methods, such as sorting for recycling, pulverization, resource recovery, and composting. These more sophisticated facilities have been renamed "materials recovery facilities."

They appear as unremarkable large hangars, most readily identified by the stream of waste collection vehicles arriving and leaving, followed by larger articulated trucks pulling covered containers.

Commercial garbage transfer stations can also accommodate the larger materials collected in skips, often from building and demolition sites. Thanks to the materials being handled, these sites are dusty. Therefore, a well-managed, modern waste transfer station will have wheel washing facilities to stop the large number of trucks moving in and out soiling the surrounding roads and causing a nuisance. In order to minimize their impact on the surrounding area, most of the sorting activities will be carried out in large covered warehouses, where loaders sort the waste into various types of materials. Many transfer stations will separate out recyclable materials such as plastics, metal, and wood.

GAR-BARGE

In April 1987, a barge containing 3,492 tons (3,168 tonnes) of garbage set sail from New York, headed for North Carolina, where, after hearing rumors that the load contained medical waste, the barge was denied permission to unload. As a result, the barge traveled down the east coast of the United States, searching for a place to unload. It sailed as far south as Belize before returning to New York, where, after a brief legal battle, the cargo was incinerated. A change in environmental legislation, coupled with extensive press coverage, led to the false perception that American landfills were full. Public interest in the disposal and recycling of our waste increased and alternative solutions to the problem of waste disposal were subsequently implemented.

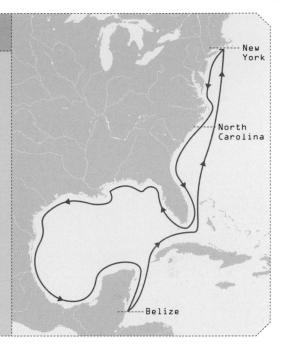

Waste collection in the Italian city of Florence.

LANDFILL SITE

WHAT?

The oldest and most common form of waste disposal, garbage is simply piled up and buried. The kind of society in which we live is reflected in the things we throw away.

WHERE?

Landfill sites are typically located outside of towns and cities, away from residential areas.

DIMENSIONS

The world's largest landfill is Fresh Kills on Staten Island, which covers 2,200 acres (8.9 km²). New York's dumping ground between 1947 and 2001, it is now being transformed into a park.

Landfill engineering is a specialist, essential, but unglamorous branch of the engineering profession. The waste we send to landfill is hazardous, bulky, and a risk to the environment. But what appears to just be a hole in the ground is actually a carefully considered waste solution that takes into account the danger to waterways and seas, air, and health.

A landfill site doesn't treat waste, it merely compacts and contains it. Many of the chemicals present in modern consumer products are hazardous to the environment and do not safely decompose. Day-to-day operations aim to ensure that our waste is securely contained so that, in the longer term, nothing harmful escapes.

Before choosing a landfill site, the geology of the surrounding area will be examined carefully. Ideally, the land should have a solid, impermeable foundation that will support the weight of the garbage and prevent contamination of local groundwater and aquifers by leakage. This applies both during landfill usage and after closure. Before waste is deposited, an impermeable textile is laid down to create a kind of seal protecting the soil beneath.

Careful consideration is given to the number of large trucks that collect and deposit at the site and specialized diggers are used to move and compact the trash. At the working face, the compacted trash is covered by soil, chipped wood, or a temporary blanket in order to minimize odor and the attention of rats and seagulls attracted by the trash.

Trash Air

"Trash air" is a mixture of gases that contains mostly methane generated by the breakdown of the organic elements of our waste. As well as being potentially explosive, it is also hazardous to health. In large amounts, it can be collected and used to generate electricity in a gas-fired power plant, vented straight to the atmosphere, or flared-off.

HISTORIC LANDFILL

As the range of materials we use in our everyday lives increases, so the composition of our waste becomes more varied. Roman landfill sites typically contained little apart from pottery shards while a medieval "midden" may have had very small amounts of metal. Today, the range of materials used is vast.

The first modern, sanitary landfill in the USA opened in 1937, where trenching, compacting, and the daily covering of waste with soil were carried out at The Fresno Municipal Sanitary Landfill in California. Recognizing the significance of waste disposal in modern society, it has now been declared a National Historic Landmark. Archaeologists continue to dig through landfill sites of the last 2,000 years, looking for clues as to how earlier civilizations lived.

A bulldozer works the garbage heap at a landfill site in Kiev, Ukraine.

INCINERATOR

WHAT?

The first waste disposal incinerators were aptly known as "destructors." They took organic waste and set fire to it, converting it into ash, flue gas, and heat.

WHERE?

In parts of Europe, incinerators win awards for their dramatic architecture. However, your local plant may prefer a more low key approach. Currently there are 400 plants across Europe, disposing of waste and creating electricity and heat for 12 million people.

DIMENSIONS

A large building with a chimney, visited by a large number of vehicles bringing municipal waste to the incinerator for processing.

The practice of burning waste has been around for hundreds of years. Before plastics became common, a large part of domestic waste was burned on household fires. Incinerators have gone in and out of favor as methods of waste control depending on legislation and the availability of landfill sites. They are steadily becoming more popular, with new technology cleansing the flue gases of potentially harmful pollutants and using the heat provided by combustion to generate electricity.

A waste incinerator in Osaka, Japan.

Technology has improved since the first UK "destructor" was built in Nottingham in 1874. In those days, incinerators were simply a furnace with a chimney into which everything was thrown, with the incombustible "clinker," or ash, falling into an ash pit. While the basic operation remains, modern incinerators must first sort the waste and separate recyclable and toxic items. They must operate at extremely high temperatures to ensure the complete breakdown of organic substances, and must use pollution scrubbers to minimize dust and toxin release into the atmosphere. Aged incinerators that don't meet these needs lose their operating license and are usually demolished.

WASTE

NOT IN MY BACKYARD

Incinerators are easy to spot due to their large prominent chimneys and constant flow of trucks laden with waste. This may help explain why they are so unpopular with large sections of the public. In many parts of the world incinerators have a bad name, being accused of causing a wide range of health problems by releasing heavy metals and toxins into the atmosphere. While we have to do something with our waste, no one seems to want an incinerator in their backyard.

Applications to build new incinerators are often fiercely contested by local residents and support networks exist to help fight such proposals. In many cases these protests are successful, with applications being moved elsewhere or alternative waste management methods adopted.

There are various types of incinerator plant designs but the typical choice for municipal solid waste is the moving grate incinerator. In this type, the dried and compacted waste is introduced by a mechanical arm through the throat, from where it moves down over the grate to the ash pit at the other end. The temperature inside the flue is raised to 1,500°F (815°C) for a minimum of two seconds to ensure all the toxic organic substances within the waste gases are completely broken down. The flue gases then pass through heat exchangers, to create steam, which then power electricity-producing turbine generators. Finally the gases pass through a flue gas cleaning system to remove any toxins and pollutants before being released into the atmosphere.

Incinerators will run continuously throughout the year, with just one scheduled stop for maintenance. At such high temperatures, with variable loads, they are an amazing feat of engineering.

WASTE INCINERATOR

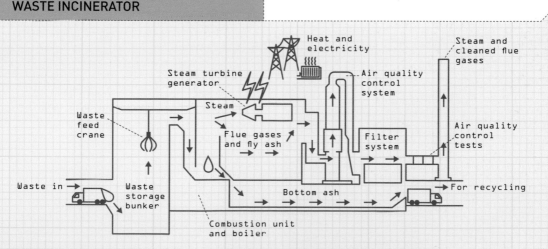

The incineration process, including waste delivery, burning, power generation, and pollution control systems. The waste is burned to produce heat, steam, ash, and flue gases. The flue gases are scrubbed and tested prior to discharge.

BIOLOGICAL REPROCESSING

WHAT?

From the humble compost heap to biodiesel production, biological reprocessing is about taking waste organic material and turning it into something useful.

WHERE?

Anywhere, from the traditional compost heap in the backyard to curbside collection for municipal anaerobic digestion. Mankind produces a lot of organic waste and the solutions for dealing with it are many and varied.

DIMENSIONS

A steaming heap about 3 ft (1 m) across in your backyard, or a series of massive cylindrical tanks at a mechanical reprocessing plant.

From farming and harvesting to eating, the process of feeding ourselves produces waste at every step. Whether the redundant leaves on the tomato plant or the scraps left on our plate, the large quantity of organic waste humans create cannot be ignored. Instead of placing it in landfill, it can be turned into heat electricity, or fuel, thereby reducing waste volumes and producing useful by-products.

Steam rising from a composting heap of organic waste.

Whether it's the compost heap in your backyard or a large-scale municipal reprocessing plant, the process is the same. The grass clippings, vegetable peelings, and other organic matter are turned into soil and fertilizer. Since over a third of our domestic rubbish is organic, this represents a significant volume that could be diverted away from the incinerator and landfill to produce a valuable but low-cost resource for local gardeners.

The Reprocessing Plant
We all know what a compost heap looks like, but a municipal biological reprocessing plant can be harder to identify as they generally don't include a steaming pile of grass cuttings.

WASTE

At such a plant, the first stage is sorting the material into organic and non-organic waste either by hand or using mechanical methods such as magnets, trommels (giant cylindrical sieves), or shredders. The non-organic waste is either sent for incineration or recycling. The organic waste is either treated with aerobic microorganisms, which break down the waste into carbon dioxide and compost, or sent to anaerobic digesters. These large enclosed cylindrical tanks mix the waste with bacteria to break down the organic material and allow safe disposal of the biogas, digestate, and water that is produced. The methane in biogas can be burned to produce both heat and electricity, usually with a standard reciprocating engine or micro-turbine. The digestate is the remains of the original material that the microbes cannot use. This brown sludge can be used for compost, low-grade building materials such as fiber-board, or feedstuff for ethanol production.

Somewhere between the compost heap and the reprocessing plant is the municipal composting facility. These are often seen at municipal recycling facilities, where garden waste is macerated and composted, then sold back to the public as soil improver.

GREASE INTERCEPTORS

With the rise in world petroleum prices, yellow grease from deep fat fryers is fast becoming yellow gold. Easily collected and processed into biodiesel, it can be used to power standard production vehicles. "Brown grease" (or "FOG"—fats, oils, and grease) from commercial kitchens is a smelly, sticky problem. Grease interceptors filter the waste water from sinks, dishwashers, and floor drains, and store the mix of food grease and chemical cleaners to prevent it entering the sewer system. With the kitchens of San Francisco alone producing over 10 million gallons of brown grease a year, schemes to turn this waste sludge into biodiesel and other fuels are being taken very seriously.

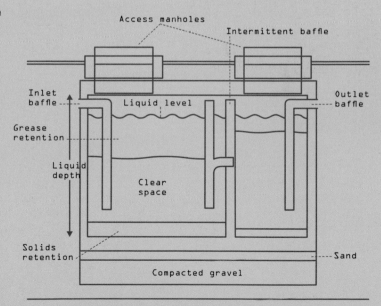

WRECKING YARD

WHAT?
Also known as a scrap, junk, salvage, or breakers yard, this is where wrecked or redundant vehicles are brought in and broken up into component parts for reuse and recycling.

WHERE?
In industrial zones on the outskirts of town, since they tend not to be the most popular neighbors and need a lot of space.

DIMENSIONS
Behind that fence will be a sizable yard. In the UK, cars are often stacked one on top of another on high metal frames to save space.

Long before modern environmental legislation and the advent of curbside recycling schemes, people were having expensive automobile accidents, wrecking cars and needing spare parts for repair. Enterprising individuals realized a long time ago that there was money to be made from the undamaged bits and pieces of our redundant cars.

The Scrap Industry
The sophistication of the wrecking yard ranges from a couple of cars half stripped in a yard corner to a warehouse with individual pieces carefully wrapped and logged into a computerized inventory.

The business model of your local second-hand spare parts dealer mirrors that of the international commodities market. When world metal prices are high, scrap yards will collect and even pay you for your wreck. When new restrictive legislation is being brought in, the economy is strong, and the demand for scrap low, then the wrecking yard will charge you.

As cars have become more sophisticated and the internet has grown, the wrecking yard has kept pace. There are still yards, known as "you pull it," where you wander around with a screwdriver and a wrench in your pocket hoping to find the right model car with just the right light cluster for your damaged vehicle before paying the yard owner. However, at most modern yards, popular and valuable items are often removed to a warehouse where, with the help of a computerized inventory, they can be plucked straight from the shelf, ready to go. Expensive engines and transmission systems may be refurbished in-house or sold on to a specialist for overhauling.

After a vehicle has been stripped of all its useful mechanical parts and trim the scrap metal is finally sold on to the specialist for compaction and melting back into raw material for reuse.

Wrecking yards may not appear to be at the cutting edge of sophisticated environmental protection solutions, but by recycling old or damaged vehicles and promoting the reuse of vehicle parts which are already manufactured they reduce pollution and resource demand.

AIRCRAFT BONE YARD

Airplanes are expensive to keep maintained and flight worthy. A last flight may be to the southwestern USA, where military and privately owned facilities are available to store planes for future use or turn them into scrap metal.

After being taken out of service, an airplane in a "bone yard" will be gradually stripped of useful parts. Expensive engines, specialized landing gear, electronics, and even munitions are removed to be recycled or stored in warehouses for the right customer. These may be sold internationally or kept in reserve pending a decision to recondition the aircraft to fly again.

Deserts with lots of space available and dry conditions to reduce corrosion are ideal locations for the aircraft bone yard.

The wreckage of a car being moved at a wrecking yard in the USA.

GLOSSARY

absolute stop: a railway signal indicating that a train must stop and cannot proceed until the signal changes.

adit: an access tunnel connected to the main tunnel of a mine.

advance signal: a railroad signal up to which a train may proceed within a block that is not completely clear.

aerial: *see* antenna

aerodrome: *see* airport

airport: a facility for aircraft takeoff and landing, for handling passengers and cargo, and servicing aircraft.

air terminal: that part of an airport providing amenities for airline passengers and administrative functions.

alternating current: electric current in which the flow constantly reverses direction. The form in which electricity is delivered to homes and businesses.

antenna: part of a radio transmitter or receiver that sends or receives radio waves.

approach signal: a railway signal warning of a restrictive signal ahead.

apron: a hard-surfaced area for parking airplanes.

aqueduct: a channel for conveying water (e.g. carrying a canal over a valley).

aquifer: a source of water underground.

arch bridge: a bridge having arches as the main supports.

arch truss: a truss having the form of an arch or arches.

at grade: at the same elevation.

bailey bridge: a bridge built of panels connected by steel pins, permitting rapid construction.

ballast: crushed stone used as a railroad bed to provide support and drainage.

bankside reservoir: for the storage of water, where the original supply may be erratic or polluted.

bascule bridge: a movable bridge consisting of balanced lifting leaves.

beam: a structural member that is longer than it is wide.

beam bridge: a bridge consisting of beams that support the roadway on their upper surface.

billboard: a large outdoor structure carrying advertising.

biological reprocessing: converting unwanted organic material into a value-added product.

biomass power: energy obtained from vegetation.

block system: a railroad system consisting of sections of track into which the entry of trains is controlled by signals.

bowstring beam: a beam or girder shaped in the form of an arch with a horizontal "string" that resists the horizontal force of loads on the arch.

boxcar: an enclosed or covered railcar.

box girder: a hollow girder or beam with a rectangular or trapezoid cross section.

cable-stayed bridge: a bridge supported by inclined cables attached to the top of a tower or at several levels up the length of the tower.

cantilever bridge: a bridge consisting of projecting spans balanced about their central point or secured at one end.

cell: a single device that produces an electrical current; a solar panel will contain several solar cells.

combined heat and power (CHP): electricity and heat distributed from the same plant.

classification yard: a railroad yard for separating trains according to car destination.

clinker: a type of black waste deposit from heating and burning processes.

cloverleaf: a grade-separated highway intersection designed in the form of a clover leaf.

coastal defenses: walls, groins, or other structures designed to protect the coastline against the action of the sea.

combined sewer: a pipe conveying storm water runoff and sewage.

conductor: a substance that allows the flow of electric current, or other form of energy.

container: a standardized reusable box used for the transport of freight.

convex: a surface that curves out at its center.

counterweight: a weight attached to a structure that balances the weight of a load in another part of the structure.

curb: a border of concrete or stones separating a sidewalk (UK: pavement) from the street.

dam: structure preventing water moving downstream.

deck: the floor or roadway of a bridge.

dike: a structure protecting land that is below sea level.

direct current: electric current that flows only in one direction; for example, a battery.

distant signal: a railroad signal at a distance from a block of track to warn that the block is closed.

dock: an area of waterway for the berthing of ships.

drawbridge: a bridge that can be raised to provide passage for ships.

dry dock: a dock from which the water can be removed to expose the bottom of a ship berthed in it.

electric charge: the property produced by adding or removing electrons to matter.

electric current: the continual flow of electrons through a wire or other electrical conductor.

electric field: the region around an electric charge.

feeder road: a minor road that feeds traffic to a major road.

fiber optics: the use of transparent fibers such as glass or plastic to transmit light.

fission: a nuclear reaction in which the nucleus of an atom splits, releasing a lot of energy.

flat car: an open, flat-decked railcar used to carry large loads and containers.

flat yard: a classification yard in which railroad cars are moved into their destination tracks by locomotives.

flight: a series of canal locks to raise boats up an incline.

floodwall: a wall that protects land from floodwater.

fossil fuels: fuels formed by the decomposition of vegetation in an environment with no oxygen.

fountain: a public source of drinking water, or a decorative feature that pumps jets of water into the air.

fractional distillation: separating a substance according to the different boiling points of its constituents.

fusion: a nuclear reaction in which atomic nuclei fuse, releasing a lot of energy.

gage: *see* track gage

garbage transfer station: a site for the storage, sorting, and transferral of waste for final disposal.

gar-barge: a boat that carries waste.

gas turbine: a turbine powered by the burning of gas.

gasometer: a pressurized storage container for natural or town gas.

geothermal energy: heat energy generated and stored within the Earth's crust.

girder: a large beam made of metal or concrete.

grade: the elevation of the surface of a road, path, railroad etc.

grade crossing: an intersection of roads, railroads, paths, or combinations of these at the same elevation.

grade separation: a grade crossing employing an underpass and overpass.

graving dock: a dry dock consisting of a basin from which water can be drained to expose a vessel's hull.

gravity yard: a classification yard in which a natural incline is used to propel railcars into their destination tracks.

groin (also groyne): a barrier built out from the shore to prevent erosion and the movement of sand and pebbles.

heat engine: an engine in which heat is converted into movement by the expansion of a gas.

home signal: a signal at the beginning of a block of railroad track that indicates whether the block is clear.

hump yard: a classification yard in which an artificial hill is used to propel railcars into their destination tracks.

hydrant: an outlet from a pipe where fluid may be drawn via a valve.

hydraulic: utilizing the mechanical properties of fluids.

hydraulic accumulator: an energy storage device for hydraulic power.

hydropower: utilizing the pull of gravity on water to provide power.

I beam: a rolled iron or steel joist having an I-shaped cross section.

incinerator: apparatus for taking the organic element of our waste and safely setting fire to it.

insulator: a substance that resists the flow of electric current, or other energy forms.

landfill: a site where waste is dumped and covered over.

levee: an embankment providing flood protection.

lift bridge: a moveable bridge whose spans are raised vertically.

lock: a chamber with gates at each end connecting two sections of a canal at different levels.

mains supply: the provision of power or services from the distributer to the consumer.

marine terminal: part of a port with facilities for docking, cargo handling, and storage.

microwaves: radio waves with a very short wavelength from 0.04 in (1 mm) to 12 in (30 cm).

mole: a breakwater of stone or concrete extending seaward from the shore.

movable bridge: a bridge that can be raised or swung aside to allow shipping to pass beneath it.

moving sidewalk: a conveyor belt for pedestrians.

nuclear power: a power plant that generates electricity using nuclear fission.

oil platform: a structure for the exploration and drilling of petrochemicals at sea.

oil refinery: a factory for the processing of petrochemical products.

overpass: the upper level of a grade-separated roadway crossing.

parking apron: a hard-surfaced area used for parking aircraft.

GLOSSARY

photovoltaic: the creation of electric power in material exposed to light.

pier: a solid foundation beneath the upright of a bridge, or a structure extending from the shore to provide berthing for ships.

pump jack: device for lifting oil using a device characterized by a distinctive counterbalanced beam.

pylon: a tower supporting the cables of a cable-stayed or suspension bridge.

quay: a structure used for loading and unloading ships.

radioactivity: the emission of radiation from disintegrating atomic nuclei.

radio waves: invisible electromagnetic waves.

railroad: a line of rails on which freight cars and passenger cars are drawn by locomotives.

reinforced concrete: concrete containing steel rods or wire mesh.

reinforcing bars (rebar): steel rods used in concrete for reinforcement.

reservoir: a natural or artificial pond or lake that holds water for onward distribution.

revetment: a sloped facing built on a shoreline to absorb the energy of the waves and reduce erosion.

riprap: irregular rock and bolders used as a seawall or revetment to protect the shoreline.

runway: a hard-surfaced strip on which aircraft land and take off.

seawall: a wall or embankment built to protect the shoreline from erosion.

service reservoir: a store of fully treated drinking water.

sleeper: a wooden or concrete beam on which railroad tracks are fixed to hold them in position and at the correct gage.

solar cell: a device which converts light into electricity.

storm drain: infrastructure for the disposal of excess rainwater.

suspension bridge: a bridge whose deck is suspended from cables running over towers.

third rail: a metal rail that carries electric current to power a locomotive, or a light rail or rapid transit car.

through arch bridge: a bridge in which the roadway is suspended from the arch rather than being carried on top of it.

track gage: the distance between the inner faces of the rails of a railroad track.

traffic signal: a device that uses lights to regulate the movement of traffic.

transformer: a device for altering the voltage of an electrical current.

transmission tower: steel lattice tower carrying power lines above ground.

truss: a structural frame composed of triangular forms whose members are variously in tension and compression.

truss bridge: a bridge composed of trusses.

turbine: a machine that produces power via blades that are turned by the movement of a liquid or a gas.

valley dammed reservoir: a dam that employs valley sides as walls.

water tower: an elevated tank that distributes water at mains pressure.

wharf: an area of waterway used by vessels to load and unload passengers and cargo.

windmill: a structure that harnesses the power of the wind via rotating sails to power machinery.

wind turbine: a device that converts the power of the wind into electricity.

REFERENCES

CHAPTER 1:
RAW MATERIALS

www.losapos.com
www.dmtcalaska.org
theenergylibrary.com
www.dimensionsguide.com
www.hebronbrick.com
www.madehow.com
www.beyondroads.com
www.tpub.com
www.clrp.cornell.edu
www.steel.org
www.thepotteries.org
www.worldsteel.org
www.uky.edu
ec.europa.eu
www.ifc.org
www.miningbasics.com
www.coalleader.com

CHAPTER 2:
WATER

Becher, Bernd and Hilla, *Basic Forms of Industrial Buildings*, Thames & Hudson, 2005

Bergeron, Louis and Maiullari-Pontois, Maria, *Industry, Architecture and Engineering: American Ingenuity, 1750-1950*, Harry N. Abrams, 2000

Chaubin, Frederic, *CCCP: Cosmic Communist Constructions Photographed*, Taschen GmbH, 2011

Gray, N.F., *Water Technology: An Introduction for Environmental Scientists and Engineers*, Elesvier, 2010

Olesen, John Peter, *The Oxford Book of Engineering and Technology in the Ancient World*, OUP, 2010

Ratnayaka et al, *Tworts Water Supply*, 6th Edition. Elesvier 2009

Turpin, Trevor, *Dam. Reaktion Books Ltd*, 2008

Vitruvius, *On Architecture*, Penguin, 2009

Watkins, George, *Stationary Steam Engines of Great Britain: The National Photographic Collection*, Landmark Collectors Library, 2003

CHAPTER 3:
POWER

The Pylon Appreciation Society, devoted to Transmission Towers www.pylons.org

Gibson, Charles Robe, *The Romance of Modern Manufacture: A Popular Account of the Marvels of Manufacturing*, London, 1910

Heald, Henrietta, *William Armstrong Magician of the North*, Northumbria Press, 2010

Hills, Richard L., *Power from Steam: A History of the Stationary Steam Engine*, Cambridge University Press, 1994

Hills, Richard L., *Power from Wind: A History of Windmill Technology*, Cambridge University Press, 1996

Margaine, Sylvain, *Forbidden Places: Exploring Our Abandoned Heritage*, Jonglez, 2009

Powell, P.E., *Windmills and Windmotors*, Lindsay Publisher, 1995

Seifried, Dieter and Witzel, Walter, *Renewable Energy: The Facts*, Earthscan, 2010

CHAPTER 4:
TRANSPORT

Road
Loukaitou-Sideris, Anastasia and Ehrenfeucht, Renia, "Vibrant Sidewalks in the United States: Re-integrating Walking and a Quintessential Social Realm" from Access Magazine, University of California Transportation Center, Number 36, Spring 2010 (www.uctc.net/access)

www.alaskaroundabouts.com
www.htma.co.uk
www.fhwa.dot.gov
www.wellington.govt.nz
whereroadsmeet.8k.com
www.dot.state.oh.us
www.kurumi.com
www.interstate-guide.com
www.cbrd.co.uk
www.ctre.iastate.edu
broadway.pennsyrr.com
history-world.org
www.direct.gov.uk
www.trafficsign.us
www.th.gov.bc.ca
www.dft.gov.uk
www.mapsofworld.com

REFERENCES

Canals

www.historyworld.net
www.canaljunction.com
www.canadiancanalsociety.org
www.jim-shead.com
www.eurocanals.com
www.falkirk-wheel.com
www.waterscape.com
www.nps.gov
www.worldcanals.com

Rail/Tunnels

www.railway-technical.com
www.mto.gov.on.ca
www.fra.dot.gov
www.mrsc.org
www.tunneltalk.com

Bridges/Ports

Koglin, Terry L., *Movable Bridge Engineering*, John Wiley and Sons, New Jersey, 2003

library.thinkquest.org
www.design-technology.org
www.forthbridges.org.uk
www.croatianhistory.net
www.nyc.gov
www.chennaiport.gov.in
www.portmetrovancouver.com

Air Transport

www.centennialofflight.gov
www.aci.aero
web.mit.edu
www.glidepathgroup.com
virtualskies.arc.nasa.gov
www.faa.gov
wiki.flightgear.org
www.nappf.com
www.pilotfriend.com
www.tpub.com

CHAPTER 5: COMMUNICATION

Barty-King, Hugh, *Girdle Round the Earth: History of "Cable and Wireless,"* William Heinemann Ltd, 1980

Camerota, Remo, *Drainspotting: Japanese Manhole Covers*, Mark Batty Publisher, 2010

Domela, Laura, *Neon Boneyard*, Lulu, 2006

Jacobs, Jane, *Death and Life of Great American Cities*, Random House, 1997

Krols, Brigit, *Sensational Billboards in Advertising*, Tectum, 2007

Macaulay, David, *The Way Things Work*, Dorling Kingsley, 2004

Melnick, Mimi, *Manhole Covers*, MIT Press, 1994

Minns, Jonathan Ellis, *Model Railway Engines (Pleasures and Treasures)*, Weidenfeld & Nicolson, 1969

CHAPTER 6: WASTE

Butler, David, *Urban Drainage*, Spons, 2004

Capes, William Parr, *Municipal Housecleaning: The Methods and Experiences of American Cities in Collecting and Disposing of Their Municipal Wastes, Ashes, Rubbish and Garbage*, 1918.

Davis, Helen, *The Archaeology of Water*, Tempus, 2008

George, Rose, *The Big Necessity: Adventures in the World of Human Waste*, Portobello Publications, 2008

Grant, Nick; Moodie, Mark and Weedon, Chris, *Sewage Solutions: Answering the Call of Nature* (third edition), Centre for Alternative Technology, 2005

Halliday, Stephen, *The Great Stink of London: Sir Joseph Bazalgette and the Cleansing of the Victorian Metropolis*, The History Press, 2001

Lambton, Lucinda, *Temples of Convenience and Chambers of Delight*, Pavilion Books, 1997

Watkins, George, Stationary Steam Engines of Great Britain: The National Photographic Collection. Landmark Collectors Library, 2003

INDEX

PHOTO CREDITS